Telecourse Student Guide

for

UNIVERSE
The Infinite Frontier

Telecourse Student Guide

for

UNIVERSE
The Infinite Frontier

Stephen P. Lattanzio

Joel M. Levine

Valerie Lynch Lee

Coast Community College District
Costa Mesa, California

Wadsworth Publishing Company
I(T)P™ An International Thomson Publishing Company

Belmont • Albany • Bonn • Boston • Cincinnati • Detroit • London • Madrid • Melbourne
Mexico City • New York • Paris • San Francisco • Singapore • Tokyo • Toronto • Washington

Coast Community College District

William M. Vega, Chancellor

Leslie Purdy, President, Coastline Community College, and Executive Producer

Jim LeMay, Administrative Dean, Office of Instructional Systems Development

David P. Stone, Producer

Valerie Lynch Lee, Instructional Designer

Michael S. Werthman, Publications Editor

Marji James, Assistant Publications Editor

Content Advisors: Stephen P. Lattanzio, Professor of Astronomy, Orange Coast College; Joel M. Levine, Professor, Physics and Astronomy, Orange Coast College.
Content Reviewer: David A. Pierce, Professor of Astronomy, El Camino College.

Published by Wadsworth Publishing Company.

Printed in the United States of America
2 3 4 5 6 7 8 9 10—01 00 99 98 97 96 95

For more information, contact Wadsworth Publishing Company,

A Division of International Thomson Publishing Inc.

I(T)P   The ITP logo is a trademark under license.

Wadsworth Publishing Company
10 Davis Drive
Belmont, California 94002 USA

International Thomson Editores
Campos Eliseos 385, Piso 7
Col. Polanco
11560 México D.F. México

International Thomson Publishing Europe
Berkshire House 168-173
High Holborn
London, WC1V 7AA England

International Thomson Publishing GmbH
Königswinterer Strasse 418
53227 Bonn Germany

Thomas Nelson Australia
102 Dodds Street
South Melbourne 3205
Victoria, Australia

International Thomson Publishing Asia
221 Henderson Road
#05-10 Henderson Building
Singapore 0315

Nelson Canada
1120 Birchmount Road
Scarborough, Ontario
Canada M1K 5G4

International Thomson Publishing Japan
Hirakawacho Kyowa Building, 3F
2-2-1 Hirakawacho
Chiyoda-ku, Tokyo 102 Japan

ISBN 0-534-20610-7

Preface

To the Student

Welcome to the telecourse *UNIVERSE: The Infinite Frontier,* an introductory course in astronomy. Whether you are planning a career in science, taking the course as part of an academic course for college credit, or taking it just for pleasure, we believe that you will find *UNIVERSE: The Infinite Frontier* interesting, entertaining, enriching, and inspiring. *UNIVERSE: The Infinite Frontier* brings you the latest astronomical discoveries and astrophysical theories. The course covers four major areas: Exploring the Sky, The Stars, The Universe of Galaxies, and Planets in Perspective. In this course you will have a breathtaking view of nature, from particles smaller than the atom to giant superclusters of galaxies. As an introductory astronomy course, *UNIVERSE: The Infinite Frontier* goes beyond describing the contents of outer space. It examines the scientific method and considers both its promises and limitations. It answers many of our questions and poses new ones time after time, so that we are continually probing into the innermost secrets of time and space.

Course Themes

The designer, academic advisors, and producers of *UNIVERSE: The Infinite Frontier* have developed this course around eight principal themes:

1. The universe is dynamic and continually evolving.

2. The universality of physical laws discovered on Earth allows us to analyze and draw conclusions about celestial phenomena that can only be studied at great distances.

3. Scientific conclusions must be based on an exacting comparison of hypotheses to evidence obtained from observations and experimental data.

4. The universe, because it has objects and environments that cannot be duplicated on Earth, is a unique laboratory for testing scientific hypotheses.

5. Because astronomy is a human endeavor, it is subject to both the limitations and enhancements of personal relationships, biases, inspiration, and creativity.

6. Most astronomical knowledge accumulates incrementally, with each new piece of knowledge providing a potential foundation for further understanding.

7. Observational and computational technologies play critical roles in shaping our understanding of the universe.

8. In addition to scientific value, astronomy has practical and philosophical value because humans are participants in, as well as observers of, the universe.

Course Components

Video Programs

As with most college courses, *UNIVERSE: The Infinite Frontier* has a text, a study guide, assignments, and tests. In addition, *UNIVERSE: The Infinite Frontier* has as a special learning element that traditional classroom-based college courses do not have: a companion half-hour video program for each of the 26 lessons in the course.

The 26 video programs for *UNIVERSE: The Infinite Frontier* feature leading practitioners, theoreticians, and academics in the fields of astronomy, planetary science, and astrophysics, who describe and explain celestial objects and events. The unique capabilities of the video will introduce you to the night sky, the life cycle of the stars, the universe of galaxies and the history of the universe, and the origin, characteristics, and evolution of the solar system.

The video programs will take you inside the data rooms of leading observatories. You will see the latest images from NASA and the Jet Propulsion Laboratories and from Earth-based telescopes, space observatories, and the Hubble Space Telescope. A special feature of the video programs is scientifically accurate 3-dimensional animation and computer graphics to illustrate complex concepts in astronomy.

Textbook

The textbook is *Horizons,: Exploring the Universe,* 1995 edition, by Michael A. Seeds (Belmont, Calif.: Wadsworth Publishing Company, 1995). The content of the textbook parallels the content of the video programs.

Telecourse Student Guide

This telecourse student guide is your road map through *UNIVERSE: The Infinite Frontier*. It is a starting point for each lesson because it contains step-by-step assignments for reading, viewing, and doing related activities, overviews of each lesson's content and the accompanying video program, and a complete array of learning activities to help you master the learning objectives for the lesson.

Each lesson in this guide has the following components:

Assignments: Detailed instructions on activities and reading assignments to be completed before and after viewing each video program.

Overview: An introduction to the main topics covered in the textbook and video program.

Learning Objectives: Statements of what you should learn from reading the textbook assignments, completing the activities in this guide, and viewing the video programs.

Viewing Notes: A capsule description of the video program for each lesson plus questions for you to consider while watching the program. The Viewing Notes are *not* a substitute for actually watching the video program.

Review Activities: Matching and Completion exercises to help you review and reinforce your understanding of important terms and concepts.

Self-Test: A brief multiple-choice quiz that allows you to test your understanding of the material in the lesson.

Using What You've Learned: Many of the lessons include suggestions for additional activities to enhance your understanding and knowledge of astronomy. (Your instructor will indicate how you are to incorporate these exercises into your studies and which ones to do, and he or she may designate them as assignments for extra credit.)

Answer Key: This section, at the end of each lesson, provides answers for the Matching and Completion items and for the Self-Test questions. Check your answers after you have completed each activity. If you have any incorrect answers, review the material.

Background Notes: In addition to the sections described above, many lessons in this student guide have accompanying Background Notes, which are assigned in addition to the text material. The Background Notes contain information that supplements material presented in the video programs and text. Your careful study of the Background Notes is essential to your achieving the Learning Objectives stated for the lesson. The Background Notes for any given Unit of this guide appear in a separate section following the last lesson in that Unit, as shown in the table of contents.

How to Take a Telecourse

If you are new to college courses in general, and to telecourses in particular, you can perhaps profit from a few suggestions offered by students who have successfully completed other telecourses.

Telecourses are designed for busy people—people with full-time jobs or family obligations—who want to take a course at home, fitting the study into their own personal schedules. To complete a telecourse successfully, you will need to plan in advance how to schedule your viewing, reading, and study. Buy the textbook, guide, and any other materials required by your instructor before the course begins and look them over; familiarize yourself with any materials supplied by your college and estimate how much time it will take you to complete special tests and assignments for each lesson. Write the dates of midterms, finals, review sessions, and special projects on your calendar so that you can plan to have extra time to prepare for them. You may find it enjoyable and instructive to watch the programs with other people, but save the talking and discussion until after the programs so that you won't miss important information. After the program, take a few minutes to write a brief summary of what you have seen and the meaning of key concepts and terms and to answer the questions listed at the end of the Viewing Notes for each lesson.

The following suggestions about how to study and how to complete *UNIVERSE: The Infinite Frontier* successfully have been compiled from students who have satisfactorily completed telecourses.

- *Do* buy both the text and the telecourse student guide for *UNIVERSE: The Infinite Frontier* or arrange to share copies with a friend. Do *not* try to get through this course without these books.

- *Do* watch each of the video programs. In order to pass the examinations, you will need to read and study the text and to view the programs. If you have a videocassette recorder, tape the programs for later review.

- *Do* keep up with your work for this course every week. Even if you do not have any class sessions on campus or any assignments to turn in, you should read the text and do the assignments in the student guide, as well as watch the video programs. Set aside viewing, reading, and study time each week and stick to your schedule.

- *Do* get in touch with the faculty member who is in charge of *UNIVERSE: The Infinite Frontier* at your college or university. The instructor can answer any questions you have about the material covered in the course. Your faculty member can also help you catch up if you are behind, advise you about additional assignments, discuss the type of test questions you can expect, and tell you where you might be able to watch programs you have missed or wish to review.

- *Do* complete all the Review Activities and Self-Tests provided in this guide. These will help you master the Learning Objectives and prepare for formal examinations.

- If you miss a program or fall behind in your study schedule, don't give up. Many television stations repeat broadcasts of the programs later in the week or on weekends. Your college might have videocassette copies of programs available in the campus library or media center. And *do* call on your course faculty member or manager to help if you have problems of any kind. This person is assigned specifically to help you succeed in *UNIVERSE: The Infinite Frontier*.

Acknowledgments

Producing the *UNIVERSE: The Infinite Frontier* telecourse was a complex team effort by many skilled people. Several of those persons responsible for this course are listed on the copyright page of this book.

In addition to those individuals, appreciation is expressed for the contributions of a number of academic advisors to the telecourse. First, Michael A. Seeds, Ph.D., professor of astronomy at Franklin and Marshall College and author of the textbook, served as a technical advisor to the series and helped in numerous ways throughout production of the video programs.

Production of *UNIVERSE: The Infinite Frontier* was made possible with grants from several educational institutions. Each of them provided a faculty advisor who helped to formulate the overall course outline. These reviewers and the institutions they represented are:

Michael A. Matkovich, M.S., Oakton Community College
(Northern Illinois Learning Resources Consortium)

William McCord, M.S., Valencia Community College
(Florida State Department of Education)

David Pierce, Ph.D., El Camino College
(INTELECOM)

The reviewers listed above were joined by six other astronomers to form a national academic advisory committee for the telecourse. The committee members proposed the telecourse themes and objectives, reviewed the video programs as they were developed, and provided innumerable valuable suggestions on content and resources for the video programs. The members of the advisory committee are listed below.

ASTRONOMY TELECOURSE
NATIONAL ACADEMIC ADVISORY COMMITTEE MEMBERS

George Jacoby, Ph.D.
  Kitt Peak National Observatory

Stephen P. Lattanzio, M.A.
  Orange Coast College

Joel M. Levine, M.A.
  Orange Coast College

Michael A. Matkovich, M.S.
  Oakton Community College

William McCord, M.S.
  Valencia Community College

Benjamin Peery, Ph.D.
  Howard University

David Pierce, Ph.D.
  El Camino College

Michael Seeds, Ph.D.
  Franklin & Marshall College

Harding Eugene Smith, Jr., Ph.D.
  University of California, San Diego

Sally Stephens, Ph.D.
  Astronomical Society of the Pacific

Appreciation is also expressed to Wadsworth Publishing Company for its support of the telecourse production.

The 26 programs of *Universe: The Infinite Frontier* were produced in the studios of KOCE-TV. An affiliate of the Public Broadcasting Service (PBS), KOCE-TV is owned and operated by the Coast Community College District. This study guide and other materials for this course were developed for publication by the Office of Instructional Systems Development at Coastline Community College (Fountain Valley, California), a member of the Coast Community College District.

Special thanks are due to those who contributed to the preparation of this study guide for publication: Dixie Deckard, Sandi Schumaker, and Thien Vu.

Contents

The Scale of the Cosmos

Assignments

For the most effective study of this lesson, we suggest that you complete the assignments in the sequence listed below.

Before Viewing the Video Program

- Read the Overview and Learning Objectives for this lesson. Use the Learning Objectives to guide your reading, viewing, and thinking.

- Read Chapter 1, "The Scale of the Cosmos," pages 3–18, in the textbook.

- Study Appendix C, "Systems of Units," pages A11–A12, in the textbook. Knowledge of this material is essential to understanding much of the content of this course. Also familiarize yourself with the tables in Appendix D, "Astronomical Data," pages A13–A24.

- Read Background Notes 1, "Scientific Method," following Lesson 6 in this guide.

- Read the Viewing Notes in this lesson.

View "The Scale of the Cosmos" Video Program

After Viewing the Video Program

- Briefly note your answers to questions listed at the end of the Viewing Notes.

- Review all reading assignments for this lesson, especially the textbook Chapter 1 summary on page 17, the Background Notes 1, and the Viewing Notes in this lesson.

- Write brief answers to all the review questions at the end of textbook Chapter 1 to be certain you understand the text material.

- Complete the Review Activities in this guide to reinforce your understanding of important terms and concepts. Check your answers with the Answer Key at the end of this lesson and review when necessary.

- Take the Self-Test in this guide to measure your achievement of the Objectives. Check your answers with the Answer Key and review when necessary.

- Complete the Using What You've Learned activities and any other activities and projects assigned by your instructor.

Overview

In a sense, astronomy is the oldest of sciences because humankind has always studied the stars. For early people on our planet, the sky and the objects moving through it were their clock, their calendar, and their compass. These celestial events ordered their lives, much as they still order our lives today. Our days are defined by the rotation of our planet; our years by the orbit of our planet around the sun. All our calendars have been based on the cycles of either the moon or the sun. Our directions are identified by the path of the sun through our sky; and the sun and the stars serve as guideposts in our travels.

The science of astronomy as we think of it today began with the ancient Greeks more than 2,000 years ago. They were the first to observe the sky systematically and to offer explanations for the movement of celestial objects based on reason and logic rather than on myth and the supernatural. But from the time of the Greeks until the invention of the telescope in 1608, astronomical observations were limited by what the naked eye could see. Since the seventeenth century, however, the story of astronomy has been one of studying a frontier that expanded with each advance in telescopes and technology. As astronomers saw more, they realized that the universe was larger, more diverse, and older. They learned that stars and clouds of gas and light were sprinkled throughout the universe. Each discovery challenged their assumptions about the universe.

Although astronomers have observed the sky for centuries, it wasn't until the middle of the nineteenth century that they first measured distances to stars. As telescopes became more powerful and measurements more precise, they learned that the universe was far larger than they had imagined: stars were trillions or thousands of trillions of miles away. A new vocabulary evolved to describe these distances. The light year was a measure of distance and a term that revealed we could not ignore the element of time in our study of the universe. The early part of the twentieth century saw more advances in our understanding of the universe. Astronomers learned how stars shine and measured the distances to "spiral nebulae." Discovery of these spiral nebulae, now known as galaxies, revealed a universe that stretched millions of trillions of miles, if not farther, and pushed the boundaries of astronomical exploration.

Throughout the twentieth century, astronomers not only pushed the frontier outward, they also began to explore it in different ways. They learned that the universe reveals itself not only in visible light, but also in the energies (light) of the infrared, radio, gamma, ultraviolet, and X ray. Astronomers placed telescopes at the tops of mountains and in orbit around Earth. Rockets put satellites into space, where they can orbit for years above Earth's atmosphere and send back data about the energies that never reach the surface of Earth. We have sent spacecraft to land humans on the moon and to explore close up all but one of the planets in our solar system. Spacecraft have orbited a few of the planets and landed on some of them. Astronomy, the oldest of sciences, has become humankind's newest science, with technology leading astronomers, almost daily, to discoveries that challenge our understanding.

The program for this lesson and the remaining lessons in this telecourse introduce to you this infinite frontier we call the universe. It will be an exciting and fascinating journey. On this journey, you will learn about things you may have never thought possible. You will stretch your imagination to visualize objects that appear only as points of light in a telescope, but, in actuality, may consist of billions of stars or reveal the most violent events in the universe. You will learn about the laws that govern the universe and some of the hypotheses and theories that may change our present knowledge. You will also learn about the people who practice astronomy and why they find it such a fascinating and important human endeavor. We are the only species on Earth that can wonder about the universe, and we are the only species on Earth that can explore the infinite frontier.

Learning Objectives

When you have completed all assignments for this lesson, you should be able to:

1. Identify the major structures of the universe and order them according to size and distance from Earth. TEXTBOOK PAGES 7–17; VIDEO PROGRAM.

2. Understand the relative distances of objects in the universe. TEXTBOOK PAGES 7–17; VIDEO PROGRAM.

3. Understand and use scientific notation. TEXTBOOK PAGES 8 and A11.

4. Define and explain the terms light year and astronomical unit. TEXTBOOK PAGES 9 and 12; VIDEO PROGRAM.

5. Explain what is meant by the scientific method and identify its principal components. BACKGROUND NOTES 1.

Viewing Notes

The first video program of *UNIVERSE: The Infinite Frontier* provides a preview of the celestial objects and events you will learn about in the course. Also, you will gain an understanding of the vast distances and the scale of the universe.

"The Scale of the Cosmos" opens with astronomers explaining why they study the stars and planets and about the excitement of the field and its importance to human knowledge. The program then goes on a visual tour of the universe, starting with the sun and moving to the very edge of the known universe. This journey, illustrated with stunning computer animation and photographs from NASA, the Jet Propulsion Laboratory (JPL), and astronomical observatories, introduces the stars, galaxies, and planets that you will learn more about later in the telecourse. As the narration describes the journey out from the sun, you will learn about the immense scale of the universe and how we define distances in terms of the astronomical unit and light year.

In the segment entitled "Vast," astronomers explain the enormous distances in the universe, providing several analogies with Earthbound objects and distances. Computer graphics reinforce the concept of the universe's immense scales of time and distance. In addition to being incredibly vast and old, the universe is richly populated with trillions of constantly evolving objects. Photographs and computer animation show some of these fascinating celestial objects: galaxies, white dwarfs, red giants, supernovae, pulsars, black holes, quasars.

This introduction to astronomy concludes by presenting the basic assumption on which the science of astronomy is based: that the physical laws we observe on Earth are the same everywhere—that the laws of nature are, indeed, universal. Astronomers describe the unique challenges of the science and how they study the stars and apply these physical laws. As Michael Seeds, author of the textbook for this course, says in this segment, astronomers "try to figure out what the rules are that describe how the universe works." Finally, several astronomers who will appear in later video programs pose some of the fundamental questions that astronomy seeks to answer about the universe. The search for answers to those questions is the true motivation for studying the infinite frontier.

As you watch the video program, consider the following questions:

1. What is the astronomical unit and when is it used most conveniently?

2. What is the light year and when is it used most conveniently.

3. How is distance related to the past history of the universe?

4. What do we mean when we say that the "laws of physics are universal"?

5. What questions do astronomers attempt to answer?

Review Activities

Matching

Match the terms listed below with the definitions that follow. Check your answers with the Answer Key and review any terms you missed.

_c_ 1. solar system   _f_ 5. Milky Way galaxy

_b_ 2. scientific notation   _a_ 6. spiral arms

_g_ 3. astronomical unit   _d_ 7. Local Group

_e_ 4. light-year

a. area of our galaxy where stars are born in great clouds of gas and dust

b. the system of recording very large or very small numbers by using powers of 10

c. the sun and its planets, asteroids, comets, and other minor objects

d. the small cluster of galaxies to which our sun belongs

e. a unit of distance; the distance light travels in one year

f. the large system of billions of stars in which our sun and solar system are located

g. average distance from Earth to the sun

Completion

Fill each blank with the most appropriate term from the list for that paragraph. A term may be used once, more than once, or not at all. Check your answers with the Answer Key and review when necessary. If a question requires two or more answers in succession, they may be in any order, unless the question indicates otherwise.

1. Our sun is a _star_ and belongs to a galaxy called the _Milky Way_. The sun is just one of billions of _stars_ that make up the galaxy. Our galaxy is one of about two dozen galaxies that make up a _Local Group_, _cluster_ known as the _Local Group_. This small group belongs to a much larger group of galaxies, known as a _super cluster_. In turn, these larger groups of galaxies are linked to form _filaments_, which may be the largest structures in the _universe_. In addition to containing so many stars, most galaxies are located millions, if not billions of _ly_ away. When astronomers study these objects, they must consider these great distances because it takes _time_ _yrs_ for the light from these distant objects to reach Earth. Thus, when we look at these objects, we do not see them as they presently are, but as they were, _millions_ or _billions_ of years ago.

billions	millions
cluster	star
filaments	stars
galaxies	supercluster
galaxy	thousands
light years	time
Local Group	universe
Milky Way	

2. Our planet, _Earth_, orbits a medium-sized _star_ that we call the _Sun_. When astronomers refer to distances in the solar system, they do so in terms of the distance between _Earth_ and the _Sun_. This distance is called an _AU_. For example, the distance from the _Sun_ to Mars is 1.5 _AU_, and the distance from the _Sun_ to Jupiter is 5.2 _AU_. These distances also imply that it takes a long _time_ for light from the sun to reach objects in our solar system. Light from the sun takes about _8 min_ to reach Earth and about _4 hrs_ to reach Neptune.

In comparison with distances within the solar system, distances to the stars outside our solar system are _thousands_ of times greater. Furthermore, the time it takes light to travel to the stars is on the order of _years_. To describe these distances, astronomers use a larger distance unit called a _l y_, which is defined as the _distance_ light travels in one _year_. This measurement is equal to about _63 AU_. The star nearest to our sun is _Proxima Centauri_. It is at a distance of 4.2 _lys_. This means that light, leaving this star today, will take 4.2 _years_ to reach Earth.

4 hours	astronomical units	Proxima Centauri
4 minutes	distance	Saturn
8 hours	Earth	star
8 minutes	Earth-sun	sun
33,000 astronomical units	hours	thousands
	light year	time
63,000 astronomical units	light years	year
astronomical unit	minutes	years

Self-Test

Select the one best answer for each question. Check your answers with the Answer Key and review when necessary.

1. Of the following objects, the smallest one is
   a. a star.
   b. a galaxy.
   c. a planet.
   d. the universe.

2. Which of the following objects is the most distant from Earth?
   a. Andromeda galaxy
   b. Pluto
   c. Milky Way galaxy
   d. sun

3. The distance between the sun and the nearest star is about
   a. 260 AU.
   b. 260,000 AU.
   c. 2.6 million AU.
   d. 2.6 billion AU.

4. Light takes approximately 8 minutes to travel from the sun to Earth. To reach Jupiter, it takes light from the sun about
   a. half as long.
   b. 100 times longer.
   c. 1,000,000 times longer.
   d. five times longer.

5. The most convenient method by which to write extremely large or extremely small numbers is called
   a. scientific method.
   b. notation of the scientific.
   c. scientific notation.
   d. numerical analysis.

6. A shorthand way of writing 425,000,000 is
   a. 4.25×10^8.
   b. 425×10^6.
   c. 0.425×10^9.
   d. all of the above.

7. A light year is
   a. the time it takes Earth to travel around the sun once.
   b. the distance Earth travels in the time it takes light from the sun to reach Earth.
   c. the distance light travels in one year.
   d. equal to 1 million astronomical units.

8. The average Earth-sun distance is referred to as
   a. an astronomical unit.
   b. a light year.
   c. a kilometer.
   d. a solar mile.

9. The first step in applying the scientific method to the understanding of an event or phenomenon is to
   a. develop a hypothesis.
   b. make observations.
   c. conduct experiments.
   d. develop a theory.

10. When scientists propose a hypothesis to explain a particular event or phenomenon, the hypothesis must be worded in such a way that

    a. it can be proved true.

    b. experiments can be conducted to test the hypothesis.

    c. a theory can evolve from it.

    d. it cannot be contradicted.

Using What You've Learned

According to your own interests or your instructor's assignment, complete one or more of the following activities.

1. While you are taking *UNIVERSE: The Infinite Frontier*, save copies of newspaper and magazine articles on astronomical research and events. Compare the new knowledge gained with material presented in the textbook. Carefully note any new and updated information. Organize the articles into two groups. The first group should contain articles on new and, as of yet, unexplained observations and discoveries; the second group should contain articles that describe observations that support previously proposed theories and hypothesis.

2. During the course of *UNIVERSE: The Infinite Frontier*, develop a bibliography of books and magazines that provide additional information on the topics presented in the video programs and in the textbook. Organize the bibliography around the following major topics: history, instrumentation (telescopes and satellites), stars, galaxies (normal and active), cosmology, the solar system and planets, the search for extraterrestrial life.

3. Construct a scale model of the solar neighborhood, the Milky Way galaxy, and the Local Group. Begin by establishing the scale of 1 light year per meter. The nearest star, Proxima Centauri will be a little over 4 meters (4 light years) away on the model. Calculate the distance to the nearest edge of the Milky Way Galaxy. The Galaxy is about 100,000 light years across and our solar system is located about two-thirds of the way out from the center. That puts the sun about 30,000 light years from the center of the galaxy, or 20,000 light years from the nearest edge. Determine this distance first in meters and then convert to kilometers (1,000 meters = 1 kilometer).

   Finally, calculate the scale distance to the edge of the Local Group. The Local Group is 6 million light years in diameter. Our own galaxy is about 2.5 million light years away from another large galaxy called Andromeda. Calculate the distance to the Andromeda Galaxy in our scale (1 meter = 1 light year). Convert that to kilometers.

If you are really ambitious, acquire a map of your state and, using the scale of the map, place the nearest star, the edge of the Milky Way, and the Andromeda Galaxy on the map. Start at the town in which you live and go outward from there. For example, the edge of the Milky Way is 20,000 light years from the sun, which, according to our scale, is 20,000 meters from the sun. This distance corresponds to 20 kilometers. Using your map's scale, locate the distance of 20 kilometers from your town. [If you do not understand metric distances, 1 kilometer is equal to 0.6214 mile. Therefore, 20 kilometers is about 12.4 miles. The table on page A12 in the textbook lists metric units and conversion factors.]

Answer Key

Matching

1. c
2. b
3. g
4. e

5. f
6. a
7. d

Completion

1. star, Milky Way, stars, cluster, Local Group, supercluster, filaments, universe, light years, time, millions, billions

2. Earth, star, sun, Earth, sun, astronomical unit, sun, astronomical units, sun, astronomical units, time, 8 minutes, 4 hours

   thousands, years, light year, distance, year, 63,000 astronomical units, Proxima Centauri, light years, years.

Self-Test

(Page numbers refer to the textbook.)

1. c (Objective 1; pages 7–8; video program)
2. a (Objective 1; page 16; video program)
3. b (Objective 2; page 12)
4. d (Objective 2; pages 10, A19)
5. c (Objective 3; page 8)
6. d (Objective 3; pages 8, A11)
7. c (Objective 4; page 12; video program)
8. a (Objective 4; page 9; video program)
9. b (Objective 5; Background Notes 1)
10. b (Objective 5; Background Notes 1)

2

The Sky

Assignments

For the most effective study of this lesson, we suggest that you complete the assignments in the sequence listed below.

Before Viewing the Video Program

- Read the Overview and Learning Objectives for this lesson. Use the Learning Objectives to guide your reading, viewing, and thinking.

- Read Background Notes 2A, "The Roots of Astronomy," and Background Notes 2B, "The Foucault Pendulum," following Lesson 6 in this guide.

- Read Chapter 2, "The Sky," pages 19–30, and Chapter 3, "Cycles of the Sky," section 3–1, pages 31–37, in the textbook.

- Read the Viewing Notes in this lesson.

View "The Sky" Video Program

After Viewing the Video Program

- Briefly note your answers to questions listed at the end of the Viewing Notes.

- Review all reading assignments for this lesson, especially the textbook Chapter 2 summary on page 29, the Background Notes 2A and 2B, and the Viewing Notes in this lesson.

- Write brief answers to all the review questions at the end of textbook Chapter 2 and to review questions 1–4 on page 55 to be certain you understand the text material.

- Complete the Review Activities in this guide to reinforce your understanding of important terms and concepts. Check your answers with the Answer Key at the end of this lesson and review when necessary.

- Take the Self-Test in this guide to measure your achievement of the Objectives. Check your answers with the Answer Key and review when necessary.

- Complete the Using What You've Learned activities and any other activities and projects assigned by your instructor.

Overview

This lesson begins, and the following three lessons continue, the study of the development of modern astronomy. Our current knowledge of the universe was not acquired overnight. The journey toward understanding our universe began thousands of years ago when ancient civilizations first looked upward into the sky and tried to relate what they saw there to their lives on Earth. Early humans may have been bewildered by the sky or even fearful of some phenomena. Some cultures "explained" the sky with complex mythologies. Many ancient civilizations and traditional peoples viewed the sky as a place where the sacred resides and as a source of forces that influenced their daily lives.

This lesson starts with the view of the sky as seen by ancient civilizations and traditional peoples. For most ancient peoples, observations of the sky were made for sacred, as well as practical, motives. The sky was part of the supernatural, a place that had been created by supernatural beings and where gods, goddesses, and other revered creatures lived.

Although many civilizations saw sacred meaning in the sky, they also recognized patterns and cycles that existed in the heavens. The same constellations were visible night after night, and different groups of constellations were visible during each season. They made connections between phenomena in the sky and conditions on Earth. They also connected the position of the rising sun and the height of the sun as it moved across the sky with the seasons. Once they recognized these recurring cycles, many ancient peoples developed calendars based on events occurring in the sky.

Each civilization, in its own special way, brought the sky into the daily lives of its people. They named constellations and created mythologies to explain the figures they saw in the sky. Some ancient civilizations, however, made outstanding astronomical observations. The Egyptians, for example, realized that not only did Earth experience a daily and yearly cycle, but also a much longer cycle associated with the drift of the North Star. (By the sixteenth century, astronomers recognized that this cycle was due to the precession of Earth's axis.) Other societies built massive structures that may have been used to pinpoint special astronomical sightings. Stonehenge in England, the Big Horn Medicine Wheel in Wyoming, and the pyramids in Egypt are just a few of the ancient observatories mentioned in this lesson.

Although many ancient peoples observed the sky and developed calendars, named constellations, and learned when the seasons would change, they did not develop models or propose theories to explain the phenomena they had observed. They simply believed celestial events and objects were the work of supernatural powers. Thus, these early observations were not based on scientific objectivity and reasoning.

The first break with the traditions of these early astronomers was made by the Greeks, starting around 500 B.C. They not only observed the sky, they used abstract reasoning to develop models and theories to understand the sky and explain what they had seen. One of these Greek astronomers was Hipparchus (second century B.C.), who is featured in this lesson. It was Hipparchus who, after observing the variations in brightness among the stars, devised a stellar classification system. Although he did not explain why the stars varied in brightness, the system he developed (which, with modification, is still in use today) provided a structured framework for studying the stars and objective standards for sorting them into categories. Other Greek astronomers developed the concept of the celestial sphere to explain the perceived motions of objects in the sky. In short, the ancient Greek astronomers were the first persons to superimpose an organized system on the universe.

Learning Objectives

When you have completed all assignments for this lesson, you should be able to:

1. Explain the role of constellations in ancient and modern observations of the sky and name several major constellations. TEXTBOOK PAGES 19–22; VIDEO PROGRAM.

2. Explain the concepts of celestial sphere and celestial equator. TEXTBOOK PAGES 24–26; VIDEO PROGRAM.

3. Define archaeoastronomy and describe the astronomical observations made by ancient cultures and the astronomical orientations of Stonehenge and other structures built by ancient cultures, including those of the Plains Indians and Mayas. BACKGROUND NOTES 2A; VIDEO PROGRAM.

4. Describe the six different brightness classes Hipparchus developed and explain how they became the basis for today's modern stellar magnitude scale. TEXTBOOK PAGES 22–24; VIDEO PROGRAM.

5. Describe the tilt of Earth's axis with respect to the ecliptic and explain why this tilt causes the seasons. TEXTBOOK PAGES 32–36; VIDEO PROGRAM.

6. Define and identify on a drawing the following: celestial equator, north celestial pole, south celestial pole, ecliptic, vernal equinox, autumnal equinox, summer solstice, winter solstice, sunrise, and sunset. TEXTBOOK PAGES 26–28 and 32–36; VIDEO PROGRAM.

7. Describe the phenomenon of precession and explain its causes. TEXTBOOK PAGE 34; VIDEO PROGRAM.

8. Describe the evidence that demonstrates that Earth rotates on its axis. BACKGROUND NOTES 2B; VIDEO PROGRAM.

Viewing Notes

"The Sky" explores how different cultures throughout the world over the span of human history have viewed celestial events and objects.

The first half of the video program focuses on ancient sky watchers. Archaeoastronomer Edwin Krupp describes how early cultures used their observations of the sky to organize their lives and he explains the symbolic meanings they attached to what they saw in the sky. Krupp and other astronomers compare how Navajo and Chinese cultures, for example, named such well-known constellations as the Big Dipper and Scorpius, and they describe the reverence held for the sun by several societies, including those in ancient Egypt and Mesopotamia.

This segment also features footage of two "sacred places"—Stonehenge in England and the Big Horn Medicine Wheel in northern Wyoming. These structures, along with many others throughout the world, were built not only to observe but also to honor celestial objects.

The next segment examines how the ancient Greeks brought a new perspective to the sky and, in doing so, provided the foundation for modern science. Hipparchus' stellar magnitude scale is explained, and computer graphics illustrate the six stellar magnitudes on the scale. The Greeks also developed the first model of the universe, and the video program shows how they used a celestial sphere to explain the observed motions in the sky and developed the concepts of the celestial equator and ecliptic.

The concluding segment explains the sky as we understand it today and shows how the two major cycles we live with—the day and the year—are determined by the *apparent* motion of the sun. In reality, it is Earth's motions of rotation and orbital revolution around the sun that make our sun and other stars appear to "move" across our sky. Of course, the moon and the planets of our solar system do move across our sky, because they follow their own orbits, independent of Earth's.

Finally, the video program graphically illustrates the changing position of the sun in the sky with time-lapse photography of the sun's path on four different days: the summer solstice, the autumnal equinox, the winter solstice, and the vernal equinox. Computer graphics demonstrate how the tilt of Earth's axis with respect to the sun causes our seasons and how precession causes the location of constellations as seen from Earth to change over time.

As you watch the video program, consider the following questions:

1. What were the reasons ancient civilizations observed the skies and how were these observations incorporated into the daily lives of the people?

2. What are Stonehenge and the Big Horn Medicine Wheel?

3. What is the magnitude scale of Hipparchus?

4. What is the difference between the celestial equator and the ecliptic?

5. How and why do seasons occur? What is meant by the terms equinox and solstice?

6. What is precession and how does it affect our observations of the night sky?

Review Activities

Matching

Match the terms listed below with the definitions that follow. Check your answers with the Answer Key and review any terms you missed.

_C_ 1. constellation

_G_ 2. asterism

_q_ 3. magnitude scale

_f_ 4. apparent visual magnitude

_j_ 5. celestial sphere

_n_ 6. north celestial pole

_e_ 7. south celestial pole

_a_ 8. ecliptic

_L_ 9. vernal equinox

_P_ 10. autumnal equinox

_I_ 11. summer solstice

_b_ 12. winter solstice

_K_ 13. precession

_O_ 14. perihelion

_h_ 15. aphelion

_m_ 16. evening star

_d_ 17. morning star

_S_ 18. rotation

_U_ 19. minute of arc

_t_ 20. celestial equator

_V_ 21. revolution

_R_ 22. second of arc

a. apparent path of the sun around the sky

b. point on the celestial sphere where the sun is farthest south

c. a stellar pattern identified by name, usually of mythological gods, people, or animals

d. any planet visible in the sky just before sunrise

e. point on the celestial sphere directly above Earth's South Pole

f. the brightness of a star as seen by human eyes on Earth

g. a named group of stars that is not one of the traditional constellations

h. orbital point at which a planet is farthest from the sun

i. point on the celestial sphere where the sun is at its most northerly point

j. an imaginary sphere of very large radius surrounding Earth and to which the planets, stars, sun, and moon seem to be attached

k. the slow change in direction of Earth's axis of rotation

l. the place on the celestial sphere where the sun crosses the celestial equator moving northward

m. any planet visible in the evening sky just after sunset

n. the point on the celestial sphere directly above Earth's North Pole

o. orbital point at which a planet is closest to the sun

p. the point on the celestial sphere where the sun crosses the celestial equator going southward

q. system for describing the brightness of a star

r. an angular measure—each minute has 60"

s. movement of an object around its axis

t. imaginary line around the sky directly above Earth's equator

u. an angular measure—each degree has 60'

v. orbiting movement around a point outside the orbiting body

Completion

Fill each blank with the most appropriate term from the list for that paragraph. A term may be used once, more than once, or not at all. Check your answers with the Answer Key and review when necessary. If a question requires two or more answers in succession, they may be in any order, unless the question indicates otherwise.

1. The ancient Greeks and others thought that Earth was surrounded by a great, hollow, crystalline globe with the stars attached to the inside. Although they were incorrect, their model is useful to us when we describe the perceived positions and motions of objects in the sky. This model of the sky is called the _celestial sphere_. If we draw a great circle on this sphere, directly above Earth's equator, we split this crystalline sphere into two equal halves. This circle is called the _celestial equator_. If we watch the stars on this sphere, they seem to move around a point on the sky directly above Earth's North Pole, a point called the _N. celestial pole_. A corresponding point, the _S. celestial pole_ can be found in the southern hemisphere directly above Earth's South Pole. If we keep watch on the motion of the planets, the moon, and especially the sun, we notice that these objects move in a very narrow region of the sky. We define the path the sun appears to follow within this region of the sky as the _ecliptic_.

celestial equator ✓   north celestial pole ✓

celestial sphere ✓   south celestial pole ✓

ecliptic ✓

2. Seasons occur on Earth because Earth's *axis of rotation* is
tilted with respect to the *plane of Earth's orbit*, which causes the
celestial equator and the *ecliptic* to cross at two points. These
points are referred to as the *autumnal* equinox and
vernal equinox. When the sun, in its apparent path around
Earth, is at the equinox, the lengths of *night* and
day are *equal*. These positions mark the beginnings
of the seasons of *autumn* and *spring*, respectively. On the
longest day of the year, the sun is at the *summer solstice* and reaches its
highest point in the sky. On the shortest day of the year, the
sun is at the *winter solstice* and reaches its *lowest* point in the
sky. These two positions mark the beginnings of the seasons of
summer and *winter*, respectively.

autumn	night
autumnal	plane of Earth's orbit
axis of rotation	spring
celestial equator	summer
day	summer solstice
ecliptic	tilted
equal	vernal
highest	winter
lowest	winter solstice

3. A system for sorting the stars into classes was first developed by the
Greek astronomer *Hipparchus*. He sorted the stars in the sky into
~~six~~ *six* classes according to their *brightness*. This
magnitude scale assigned the numbers ___1___ through
___6___ to the stars he could see. The *brightest* stars were
assigned the *smallest* numbers. This system was somewhat
confusing, because it made the scale seem upside down. In other
words, the brightest stars were *first* class stars and the
faintest stars were *sixth* class stars. Modern astronomers
have adopted this scale and modified it slightly. The brightness of

some stars was recalculated and determined to be brighter than the original scale classes. These stars were then assigned _negative_ numbers to describe their _brightness_. Since the original scale did not include the factor of _distance_, these _magnitudes_ are known as _apparent_. This scale tells us only how bright these stars _appear_.

apparent visual magnitudes	largest
appear	lowest
Aristotle	magnitude
brightest	magnitudes
brightness ✓	negative
faintest	one
fifth	seven
first	seventh
five	six
highest	sixth
Hipparchus ✓	

Self-Test

Select the one best answer for each question. Check your answers with the Answer Key and review when necessary.

1. Astronomers today use the names of constellations to
   a. refer to groups of stars first identified by the ancient Greeks.
   b. identify asterisms.
   c. identify a specific area of the sky.
   d. refer to only the brightest star in a constellation.

2. The principal reason ancient civilizations gave names to groups of stars in the sky was probably to

   a. predict the seasons.

   b. provide a common reference among different cultures.

   c. develop a calendar.

   d. honor gods, heroes, and animals.

3. When the cosmos is described in terms of the celestial sphere, Earth is located

   a. at the center of the sphere.

   b. one third from the edge of the sphere.

   c. beyond the limits of the sphere and in orbit around it.

   d. equidistant from the edge of the sphere as the sun.

4. The great circle that divides the celestial sphere into two equal hemispheres is called the

   a. ecliptic.

   b. celestial meridian.

   c. lines of magnitude.

   d. celestial equator.

5. The Navajo Indians described the Big Dipper and some associated stars as the

   a. celestial chariot.

   b. revolving female.

   c. revolving male.

   d. swimming ducks.

6. At Stonehenge, the sun rises over the Heelstone on the day of the

   a. summer equinox.

   b. summer solstice.

   c. winter solstice.

   d. spring equinox.

7. According to the magnitude scale established by Hipparchus, a fifth-magnitude star is

   a. fainter than a fourth-magnitude star.

   b. brighter than a fourth-magnitude star.

   c. the brightest type of star.

8. Stars brighter than a first-magnitude star are identified by

   a. high positive numbers.

   b. negative numbers.

   c. low positive numbers.

   d. letters of the Greek alphabet.

9. The ecliptic is the path the sun appears to follow during the course of a year. It also defines the plane of Earth's orbit. The axis of Earth is tilted with respect to the ecliptic plane by about

   a. 50 degrees.

   b. 12.5 degrees.

   c. 23.5 degrees.

   d. 32.5 degrees.

10. The vernal equinox, the first day of spring, occurs when the sun

    a. reaches its highest daytime position in the northern hemisphere.

    b. reaches its lowest daytime position in the southern hemisphere.

    c. crosses the celestial meridian.

    d. crosses the celestial equator.

11. The winter solstice occurs when the sun is

    a. at its lowest position at noon when viewed by northern hemisphere observers.

    b. near the full moon as viewed by southern hemisphere observers.

    c. crossing the celestial equator.

    d. rising due east and setting due west.

12. The ecliptic and celestial equator cross at an angle of about

    a. 45 degrees.

    b. 32.5 degrees.

    c. 23.5 degrees.

    d. 10.0 degrees.

13. As Earth rotates on its axis and revolves around the sun, its axis slowly describes a cone in space. This motion is called

    a. oscillation of the axis.

    b. precession.

    c. retrograde motion.

    d. revolution of the axis.

14. One of the most noticeable consequences of precession (given enough time) is the

    a. change of the pole star.

    b. change of the period of Earth's orbit.

    c. disappearance of certain constellations during the year.

    d. appearance of different stars on the celestial equator during the year.

15. The fact that Earth rotates on its axis can be demonstrated by the observation that

    a. the amplitude of swing of a pendulum decreases with time.

    b. the direction of the plane of swing of a pendulum changes.

    c. a pendulum will rotate at the same rate Earth rotates.

    d. a pendulum will stop swinging after 24 hours.

16. A device that demonstrates that Earth rotates on its axis is a

    a. celestial pendulum.

    b. Newtonian pendulum.

    c. Foucault pendulum.

    d. rotational pendulum.

Using What You've Learned

According to your own interests or your instructor's assignment, complete one or more of the following activities.

1. Observe the rising (or setting) sun once a week during the course of the semester (try to continue this observation for an entire year). On a diagram of the horizon (eastern or western) record the position of the rising or setting sun. Discuss how these observations could lead to the development of a yearly calendar.

2. Place a two- or three-foot-long stick vertically in the ground. Record the length of the shadow once an hour from sunrise to sunset. How would you determine the time of local noon? How could you use this information to determine the longest day of the year, the date of the summer solstice? How could you determine the day of the winter solstice? The Chinese and many other civilizations made these observations thousands of years ago.

3. Select an ancient civilization and conduct research into the astronomy done by that civilization. Be sure to distinguish between their mythology of star names and constellations and their observations of significant astronomical events such as solstices, equinoxes, heliacal rising of particular stars, and eclipses.

Answer Key

Matching

1.	c	12.	b
2.	g	13.	k
3.	q	14.	o
4.	f	15.	h
5.	j	16.	m
6.	n	17.	d
7.	e	18.	s
8.	a	19.	u
9.	l	20.	t
10.	p	21.	v
11.	i	22.	r

Completion

1. celestial sphere, celestial equator, north celestial pole, south celestial pole, ecliptic

2. axis of rotation, tilted, ecliptic, celestial equator, ecliptic, autumnal, vernal, night, day, equal, autumn, spring, summer solstice, highest, winter solstice, lowest, summer, winter

3. Hipparchus, six, brightness, magnitude, one, six, brightest, smallest, first, sixth, negative, brightness, distance, magnitudes, apparent visual magnitudes, appear

Self-Test

(Page numbers refer to the textbook.)

1. c   (Objective 1; page 20; video program)

2. d   (Objective 1; page 19)

3. a   (Objective 2; page 24; video program)

4. d   (Objective 2; page 26; video program)

5. c   (Objective 3; video program)

6. b   (Objective 3; Background Notes 2A; video program)

7. a   (Objective 4; page 22)

8. b   (Objective 4; page 23)

9. c   (Objective 5; page 33)

10. d   (Objective 5; page 33)

11. a   (Objective 6; page 33)

12. c   (Objective 6; page 33)

13. b   (Objective 7; page 34)

14. a   (Objective 7; page 34)

15. b   (Objective 8; Background Notes 2B; video program)

16. c   (Objective 8; Background Notes 2B; video program)

Cycles of the Sky

Assignments

For the most effective study of this lesson, we suggest that you complete the assignments in the sequence listed below.

Before Viewing the Video Program

- Read the Overview and Learning Objectives for this lesson. Use the Learning Objectives to guide your reading, viewing, and thinking.

- Read Chapter 3, "Cycles of the Sky," section 3-2, pages 37–51, in the textbook.

- Read the Viewing Notes in this lesson.

View "Cycles of the Sky" Video Program

After Viewing the Video Program

- Briefly note your answers to questions listed at the end of the Viewing Notes.

- Review all reading assignments for this lesson, especially the textbook Chapter 3 summary on page 54 and the Viewing Notes in this lesson.

- Write brief answers to review questions 6–8 at the end of textbook Chapter 3 to be certain you understand the text material.

- Complete the Review Activities in this guide to reinforce your understanding of important terms and concepts. Check your answers in the Answer Key at the end of this lesson and review when necessary.

- Take the Self-Test in this guide to measure your achievement of the Objectives. Check your answers with the Answer Key and review when necessary.

- Complete the Using What You've Learned activities and any other activities and projects assigned by your instructor.

Overview

Even a casual observer soon realizes that changes occur in the sky and on Earth. The moon changes its shape and its position against the background stars. The level of the ocean at the shoreline varies from high to low. The sun on rare occasions darkens during the daytime, and the full moon on occasion turns reddish-orange at night. Tides and eclipses are the most obvious and dramatic manifestations of change in the cycles of the sky. The serious observer will notice that these changes are periodic and, with careful study, predictable. These cycles are the topic of this lesson.

Perhaps the most obvious cycle observed from Earth is the changing shapes or phases of the moon. The cycle occurs so regularly that many cultures use the phases of the moon as a measure of time. It is interesting that one of the most common misconceptions about the origin of the moon's phases is that they are caused by a portion of Earth's shadow falling on the moon. In actuality, the phases of the moon depend on how much and which part of the lunar surface is illuminated by sunlight reflected toward Earth.

The moon is bright, and its changes are easily observed. However, other cycles are more subtle, and not everyone on Earth may be able to observe all the phenomena. Tides are one example. Only persons near a shoreline can observe the changing tides. Although two high and two low tides wash up against shorelines each day, the size of those tides depends on the shape of the ocean floor and coastline at the observer's location. In some places, the differences between high and low tides may be more than 30 feet; in other places, the difference may be only a foot or less.

However interesting the phases of the moon and the cycle of tides may be, it is lunar and solar eclipses that, throughout history, have been the most remarkable and observed celestial events. These events are so dramatic that ancient civilizations made extraordinary efforts to predict the occurrence of eclipses. Even today, astronomers take great effort to predict precisely the time and location of eclipses.

Eclipses, especially total solar eclipses, give astronomers additional opportunities to study features on and near the sun that are normally not easily observed from Earth. Accurate predictions also give nonastronomers the opportunity to prepare themselves for an adventure of a lifetime—the observation of a total solar eclipse. This impressive display by nature is free to everyone who is in the right place at the right time. Although no solar eclipses will be visible over North America through 2005, several will occur in the Western hemisphere. If someday you should be fortunate enough to observe a solar eclipse, remember: **Never look directly at the sun without using a scientifically approved filter.** These filters are usually available through optical and telescope shops, local astronomy clubs, and planetariums. Please be careful. The intense light from the sun can irreparably damage your eyes in a very short time.

Until you are fortunate enough to view a total solar eclipse, you will have to rely on personal commentaries to gain some insight into the experiences of astronomers and others who have observed a total solar eclipse. Words such as "awesome," "magnificent," "events beyond my control," and "frozen in place" have been used to describe this event. Although these descriptions seem trite and overused, they are in fact appropriate words to describe one of the most amazing natural events ever observed by humans. These feelings of awe, excitement, and curiosity are at the root of our desire to understand and explore events and phenomena within our universe.

Learning Objectives

When you have completed all assignments for this lesson, you should be able to:

1. Explain, by drawing a diagram, the phases of the moon and name these phases. TEXTBOOK PAGES 37–39; VIDEO PROGRAM.

2. Explain the difference between the sidereal and synodic periods of the moon. TEXTBOOK PAGES 37–39; VIDEO PROGRAM.

3. Describe the relationship between the time for the moon to orbit Earth once (sidereal period) and the time for the moon to rotate on its axis once. TEXTBOOK PAGES 37–38; VIDEO PROGRAM.

4. Describe how the moon and the sun produce tides on Earth and explain the effect of tidal forces on Earth's rotation and the orbital motion of the moon. TEXTBOOK PAGES 39–41; VIDEO PROGRAM.

5. Explain the necessary conditions for the occurrence of a lunar eclipse. TEXTBOOK PAGE 41; VIDEO PROGRAM.

6. Explain the necessary conditions for the occurrence of a solar eclipse and the various types of solar eclipses possible. TEXTBOOK PAGES 41–51; VIDEO PROGRAM.

Viewing Notes

"Cycles of the Sky" begins with a detailed description of the phases of the moon, accompanied by computer graphics that highlight the relative positions of Earth, the moon, and the sun during each of the phases. The video program also explains the sidereal and synodic periods of the moon and why they are different.

This knowledge of the relative positions of Earth, the sun, and the moon provides a foundation for the explanation of tides on Earth caused by differences in gravitational pull on opposite sides of Earth.

The segment on lunar eclipses describes how lunar eclipses result when the moon enters Earth's shadow and explains the differences between penumbral and umbral eclipses. An especially interesting section explains why, in a total lunar eclipse, the moon sometimes appears to be a coppery red disc.

The comparatively rare but spectacular phenomenon of a solar eclipse is the next topic covered in the video program. Astronomers explain how the unique orbital and size relationships between the Earth, the sun, and the moon make a total solar eclipse possible. Photographs and film of actual solar eclipses illustrate such phenomena as Bailey's beads and the diamond ring effect. The program also explains why eclipses do not occur every month but do occur at predictable intervals, known as the Saros Cycle.

In the final segment, three astronomers describe their personal thoughts, feelings, and experiences while observing the total solar eclipse that occurred on July 11, 1991. This segment includes footage of preparations for viewing, the progress of the eclipse itself, the changing quality of light in the area of the total eclipse, the pinhole projections of light through leaves, and the solar corona and prominences visible only during a total solar eclipse.

As you watch the video program, consider the following questions:

1. What causes the phases of the moon?

2. What are the different phases of the moon and what is the appearance of each phase?

3. What is the difference between the sidereal and synodic periods of the moon?

4. What causes the tides and what is meant by differential attraction?

5. What conditions are necessary for a lunar eclipse to occur?

6. What conditions are necessary for a solar eclipse to occur?

7. What major observations can amateurs and professionals make during a total eclipse of the sun?

Review Activities

Matching

Match the terms listed below with the definitions that follow. Check your answers with the Answer Key and review any terms you missed.

_h_ 1. sidereal period

_m_ 2. synodic period

_a_ 3. spring tide

_i_ 4. neap tide

_q_ 5. lunar eclipse

_G_ 6. penumbra

_n_ 7. umbra

_b_ 8. solar eclipse

_K_ 9. photosphere

_r_ 10. chromosphere

_f_ 11. corona

_d_ 12. prominences

_P_ 13. diamond ring effect

_J_ 14. apogee

_O_ 15. annular eclipse

_C_ 16. nodes

_S_ 17. eclipse seasons

_L_ 18. Saros cycle

_e_ 19. perigee

a. ocean tide of high amplitude that occurs at full and new moon

b. the event that occurs when the moon passes directly between Earth and the sun, blocking our view of the sun

c. the points where an object's orbit passes through the plane of Earth's orbit

d. eruptions on the solar surface visible during total solar eclipses

e. the point closest to Earth in the orbit of a body circling Earth

f. the faint outer atmosphere composed of low-density, high-temperature gas visible during a total solar eclipse

g. the portion of a shadow that is only partially shaded

h. the time a celestial body takes to turn once on its axis or revolve once around its orbit relative to the stars

i. ocean tide of low amplitude occurring at first- and third-quarter moon

j. the point farthest from Earth in the orbit of a body circling Earth

k. the bright visible surface of the sun

l. an 18-year, 11-day period after which the pattern of lunar and solar eclipses repeats

m. an object's orbital period with respect to the sun

n. the region of a shadow that is totally shaded

o. a solar eclipse in which the photosphere appears around the edge of the moon in a brilliant ring, or annulus

p. during a total solar eclipse, the momentary appearance of a spot of photosphere at the edge of the moon, producing a brilliant glare set in the silvery ring of the corona

q. the darkening of the moon when it moves through Earth's shadow

r. bright gases just above the photosphere of the sun

s. the intervals of time, approximately six months apart, during which eclipses are possible

Completion

Fill each blank with the most appropriate term from the list for that paragraph. A term may be used once, more than once, or not at all. Check your answers with the Answer Key and review when necessary.

1. A total solar eclipse is probably one of the most impressive events in all of nature. This kind of eclipse occurs when the shadow of the _Moon_ falls onto the surface of _Earth_. The darkest part of the lunar shadow is called the _umbra_, and only those individuals within this portion of the moon's shadow see the total eclipse. During the eclipse, the moon blocks out the _Photosphere_ of the sun. Once totality is reached, the sun's _corona_, or faint outer atmosphere, becomes visible. Also visible during a solar eclipse are the atmosphere immediately above the sun's visible surface, the _chromosphere_, and streamers of solar material, known as _prominences_, that stretch thousands of miles into space.

chromosphere	penumbra
corona	photosphere
Earth	prominences
moon	umbra

2. In order for a solar eclipse to occur, the moon must be at
   New phase. For a lunar eclipse to occur, the moon must be at
   full phase. These phases occur every month, yet eclipses do
   not occur every month because the moon's orbit is _tilted_ with
   respect to _Earth's_ orbit, which causes the moon's orbit to cross
   Earth's orbit twice at points called _nodes_. The moon
   must be at or near these points, and at the right phase, in order for an
   eclipse to occur. The periods of time during which eclipses are
   possible are referred to as _eclipse seasons_ These periods occur _two_
   times a year and are about
   six months apart. Furthermore, since the moon's orbit is
   elliptical and not perfectly circular, the moon may be at its greatest
   distance from Earth, or _apogee_. As a result, the moon will appear
   smaller in the sky and not be able to cover the sun completely.
   The result will be a partial exposure of the _photosphere_ of the sun in
   the form of a _ring_ or _annulus_. This type of eclipse is
   called an _annular_ eclipse.

annular	perigee
annulus	photosphere
apogee	ring
Earth's	six
eclipse seasons	smaller
elliptical	sun's
four	three
full	tilted
new	two
nodes	

Self-Test

Select the one best answer for each question. Check your answers with the Answer Key and review when necessary.

1. When the moon is located in a direct line and between Earth and sun, the phase is called
   a. full moon.
   b. last quarter.
   c. new moon.
   d. first quarter.

2. The angle formed by the sun, Earth, and moon with Earth at the vertex during the third-quarter phase of the moon is
   a. 90°.
   b. 180°.
   c. 45°.
   d. 0°.

3. The sidereal period of the moon is the length of time it takes for the moon to
   a. go through its cycle of eclipses, both lunar and solar.
   b. orbit Earth and return to the same position relative to the background stars.
   c. complete its cycle of phases.
   d. change from new to full phase.

4. The length of time it takes for the moon to complete its cycle of phases is referred to as the
   a. Saros cycle.
   b. sidereal period.
   c. synodic period.
   d. lunar month.

5. The moon rotates on its axis
   a. once every 24 hours.
   b. in the same amount of time it takes for the moon to orbit Earth.
   c. once every 24 days.
   d. in a period of time equal to the cycle of the tidal bulges on Earth.

6. Because its rotation rate is synchronized with its orbital motion, the moon
   a. causes tides on Earth every 12 hours.
   b. can, at times, block out sunlight and produce a solar eclipse.
   c. enters Earth's shadow to produce a lunar eclipse.
   d. keeps the same side pointed toward Earth at all times.

7. The bulge in Earth's oceans that occurs on the side of Earth nearest the moon results from
   a. the moon's gravitational influence being greater on the side of Earth nearest the moon than it is on the center of Earth.
   b. the moon's gravitational influence being weaker on the side of Earth nearest the moon than it is on the center of Earth.
   c. Earth's gravitational influence on the oceans being weaker than the moon's.
   d. Earth's rotation, which creates centrifugal forces.

8. The extreme tides of the month that occur when the sun's and moon's gravitational influence operate together and increase the net force of gravity on the oceans of Earth are referred to as
   a. neap tides.
   b. spring tides.
   c. lunar tides.
   d. solar tides.

9. In order for a total lunar eclipse to occur,

   a. Earth must pass through the penumbra of the moon's shadow.

   b. the moon must pass through the penumbra of Earth's shadow.

   c. the moon must pass through the umbra of Earth's shadow.

   d. Earth must pass through the umbra of the moon's shadow.

10. A lunar eclipse can occur only if the phase of the moon is

    a. new.

    b. last quarter.

    c. waxing crescent.

    d. full. Because A lunar eclips is @ night. And moon lige @ sunset And the moon sets @ Dawn (light from sun lights the whole full moon)

11. In order to observe a total solar eclipse, an individual must be located completely within the

    a. penumbra of the moon's shadow.

    b. umbra of the moon's shadow.

    c. umbra of Earth's shadow.

    d. penumbra of Earth's shadow.

12. We do not see either a solar or lunar eclipse every month because

    a. the moon's orbit is tilted with respect to Earth's orbital plane.

    b. the full moon does not occur every month.

    c. the orbit of the moon is too elongated and the moon periodically misses Earth's shadow.

    d. it is not visible from the land masses on Earth each month.

13. Twice a year, the line of nodes points toward the sun. When a full or new moon approaches the line of nodes at this time,

    a. nothing happens.

    b. a lunar eclipse is followed by a solar eclipse within two weeks.

    c. a solar eclipse is followed by a lunar eclipse within four weeks.

    d. an eclipse of one kind or another is possible.

14. Because of the elongated nature of the lunar orbit, the moon is farther from Earth at some times than at others. If a solar eclipse occurs when the moon is farthest from Earth, the moon does not cover the sun completely, but leaves a ring of sunlight around it. This eclipse is called

   a. an annular eclipse.

   b. a semi-solar eclipse.

   c. a partial eclipse.

   d. a total eclipse.

Using What You've Learned

According to your own interests or your instructor's assignment, complete one or more of the following activities.

1. Observe the moon for as many nights during an entire cycle as possible. Make note of (a) the rising and setting times, (b) the motion of the moon against the background stars, and (c) the phase of the moon.

2. Photograph the moon each night for as much of the cycle as possible. (Record the dates.) Be sure to use the same lens for all exposures. (You will need only 1/125 exposure time at 400 ISO for most of these photographs. Check your meter to be sure.) Compare the diameter of the moon in each photograph and determine the dates of apogee and perigee. (The moon will appear largest at perigee.)

3. Make a list of the sources that give information concerning the dates, times, and locations for observations of lunar and solar eclipses that will occur during the next year.

4. Research the history of eclipse observations. What mythologies exist because of eclipses? What mythologies exist that attempt to explain the phases of the moon?

5. The following table lists partial and total lunar eclipses that will be visible from the continental United States through 2000. If at all possible, observe a lunar eclipse.

Partial and Total Lunar Eclipses
Visible in the Continental United States 1995–2000

Note: Times are Eastern standard time. These times may differ slightly depending on the geographical location of the observer in the United States and the local time zone. The three times listed are (a) first contact with umbra (partial or total eclipse) or penumbra (penumbral eclipse), (b) maximum eclipse, (c) last contact with umbra or penumbra.

DATE AND TIME	TYPE	VISIBILITY
1995, April 15 6:42 A.M. 7:19 A.M. 7:55 A.M.	Partial	Central, West: ends near moonset.
1995, October 8 8:58 A.M. 11:01 A.M. 1:04 P.M.	Penumbral	Northwest to central California.
1996, April 4 5:20 P.M. 7:10 P.M. 8:59 P.M.	Total	East: partial at moonrise. Central: totality in progress at moonrise. West: end of totality, partial begins at moonrise.
1996, September 26 8:10 P.M. 9:52 P.M. 11:33 P.M.	Total	Eastern half: all. Western half: partial begins at moonrise.
1997, March 23 9:57 P.M. 11:39 P.M. 1:21 A.M.	Partial	All of United States.
1998, March 12 9:14 P.M. 11:20 P.M. 1:26 A.M.	Penumbral	All of United States.

Partial and Total Lunar Eclipses
Visible in the Continental United States 1995–2000 (continued)

DATE AND TIME	TYPE	VISIBILITY
1998, August 7 8:32 P.M. 9:24 P.M. 10:17 P.M.	Penumbral	East: all. Central, West; partial begins at moonrise.
1998, September 6 4:14 A.M. 6:10 A.M. 8:06 A.M.	Penumbral	East, Central; ends at moonset. West: all.
1999, January 31 9:04 A.M. 11:17 A.M. 3:30 P.M.	Penumbral	Central, West: partial begins at moonset.
1999, July 28 3:56 A.M. 6:33 A.M. 7:45 A.M.	Partial	East, Central; partial in progress at moonrise. West, ends at moonrise.
2000, January 20 10:01 P.M. 11:44 P.M. 1:26 P.M.	Total	All of United States.
2000, July 16 6:57 A.M. 8:55 A.M. 10:54 A.M.	Total	Central, West: partial begins at moonrise.

Answer Key

Matching

1.	h	8.	b	15.	o
2.	m	9.	k	16.	c
3.	a	10.	r	17.	s
4.	i	11.	f	18.	l
5.	q	12.	d	19.	e
6.	g	13.	p		
7.	n	14.	j		

Completion

1. moon, Earth, umbra, photosphere, corona, chromosphere, prominences

2. new, full, tilted, Earth's, Earth's, nodes, eclipse seasons, two, six, elliptical, apogee, smaller, photosphere, ring, annulus, annular

Self-Test

(Page numbers refer to the textbook.)

1. c (Objective 1; page 38; video program)

2. a (Objective 1; page 38; video program)

3. b (Objective 2; page 37; video program)

4. c (Objective 2; page 38; video program)

5. b (Objective 3; page 37)

6. d (Objective 3; page 37; video program)

7. a (Objective 4; pages 39–40; video program)

8. b (Objective 4; page 40; video program)

9. c (Objective 5; page 41; video program)

10. d (Objective 5; page 41; video program)

11. b (Objective 6; page 41; video program)

12. a (Objective 6; pages 50–51; video program)

13. d (Objective 6; pages 50–51; video program)

14. a (Objective 6; page 47; video program)

The Origin of Modern Astronomy

Assignments

For the most effective study of this lesson, we suggest that you complete the assignments in the sequence listed below.

Before Viewing the Video Program

- Read the Overview and Learning Objectives for this lesson. Use the Learning Objectives to guide your reading, viewing, and thinking.

- Read Background Notes 4, "Eratosthenes' Calculation of Earth's Circumference," following Lesson 6 in this guide.

- Read Chapter 4, "The Origin of Modern Astronomy," sections 4–1 through 4–5, pages 57–74 in the textbook.

- Read the Viewing Notes in this lesson.

View "The Origin of Modern Astronomy" Video Program

After Viewing the Video Program

- Briefly note your answers to questions listed at the end of the Viewing Notes.

- Review all reading assignments for this lesson, especially the textbook Chapter 4 summary on pages 77–78, the Background Notes 4, and the Viewing Notes in this lesson.

- Write brief answers to review questions 1–9 at the end of textbook Chapter 4 to be certain you understand the text material.

- Complete the Review Activities in this guide to reinforce your understanding of important terms and concepts. Check your answers with the Answer Key at the end of this lesson and review when necessary.

- Take the Self-Test in this guide to measure your achievement of the Objectives. Check your answers with the Answer Key and review when necessary.

- Complete the Using What You've Learned activities and any other activities and projects assigned by your instructor.

Overview

This lesson surveys our understanding of the solar system from the time of the ancient Greeks to the time of Johannes Kepler. A major change in the way humans described nature began with the Greeks more than 2,000 years ago. Before that time, supernatural beings and events were used to explain the universe. In ancient Greece, such thinkers as Pythagoras, Aristotle, Ptolemy, and Eratosthenes began to use scientific reasoning and mathematics to develop models and hypotheses based on observations. In other words, science began to offer explanations based on reality rather than on divine forces. Although many of their hypotheses were incorrect, the contributions of the Greeks mark the beginning of science as we know it today.

Among the ancient Greeks, Pythagoras (c. 580–c. 500 B.C.), a mathematician and philosopher, was the first to develop the idea that Earth is a sphere (possibly based on observations of the round shadow cast on the moon by Earth during eclipses and of how a ship's mast would gradually disappear as the ship sailed over the horizon). In Pythagoras's model of the universe, Earth was surrounded by several crystalline spheres that rotated and to which the moon, sun, planets, and stars were attached.

Aristotle, 150 years after Pythagoras, refined the Pythagorean concept of the universe into a more detailed model, which served as the foundation for Western civilization's understanding of the universe for almost 2,000 years. In Aristotle's model, Earth was a stationary sphere, encircled by more than 50 crystalline spheres carrying celestial bodies. The spheres rotated westward each day, which, according to Aristotle's model, accounted for the movements of the moon, sun, planets, and stars.

To the philosophers and people of that time, Aristotle's model seemed reasonable. To them, as to us, Earth did not feel as if it were moving. Other "evidence" that Earth did not move was provided by the observation that birds and other airborne objects were not left behind as Earth moved beneath them. For the Greeks, Earth simply did not move.

Another ancient Greek astronomer who employed mathematics and scientific reasoning to learn about Earth was Eratosthenes. Around 200 B.C. he calculated the size of Earth. The method he used is described in Background Notes 4, "Eratosthenes' Calculation of Earth's Circumference."

Because explanations of how events occurred have to be supported by actual observations, Aristotle's model was found to be inadequate for understanding some celestial phenomena, such as the occasional apparent backward motions of the planets. Other Greek astronomers modified Aristotle's model until it became very complicated. Then about A.D. 140, some 500 years after Aristotle developed his model, an astronomer named Ptolemy (Claudius Ptolemaeus) created an even more sophisticated and accurate model of the universe. In Ptolemy's model, Earth was at the center of the universe and was surrounded by a perfect heaven in which planets moved at uniform speed in absolutely circular orbits around Earth. Ptolemy's model required dozens of epicycles and deferents, circles upon circles, to account for the observed motions of planets and astronomical cycles.

It was not until the middle of the sixteenth century that a major revolution occurred in the way Western civilization looked at the solar system and universe. Nicholaus Copernicus maintained that the sun was at the center of the universe. In his model, the planets and Earth moved around the sun in circular orbits. However, the evidence he was able to offer to support his hypothesis was not very convincing. It did not predict the positions of planets more accurately than did the Ptolemaic model nor did it explain the parallax shift in the stars as Earth orbited the sun. What eventually made the heliocentric, or Copernican, model so desirable was that it offered a simple, even elegant, view of nature. Copernicus and other supporters of the heliocentric model concluded that nature did things in a way that is pleasing to both the eye and intellect. The task is to explain nature in the simplest way.

Evidence was still required to support the Copernican (heliocentric) model. It was not until Galileo used a telescope to observe the night sky that such evidence was obtained. The phases of Venus and the discovery of satellites around Jupiter were the convincing observations that proved that celestial objects did not orbit Earth. Even these observations, however, were unable to make the Copernican model more accurate. It took the naked-eye observations of Tycho Brahe (before telescopes were invented) and the mathematical genius of Johannes Kepler to create a set of laws of the solar system that correctly predicted the positions of the planets. Kepler's three laws of planetary motion placed the sun not at the center of a circle but at one of the foci of an ellipse and showed the relationships between the speed of a planet, the period of its orbit, and its average distance from the sun.

Science had come a long way from the time of the ancient Greeks to the time of Johannes Kepler. Models of the universe had to be in agreement with observational evidence. Mathematical rules had to be consistent. However, science still lacked fundamental laws to explain not only what happened in the heavens, but also events that occurred on Earth. Pieces of a puzzle were beginning to fit together. Science was on the threshold of a major advance. That advance was started by Isaac Newton and continued by Albert Einstein. Their contributions to our understanding of the universe are topics covered in the next lesson.

Learning Objectives

When you have completed all assignments for this lesson, you should be able to:

1. Describe how Eratosthenes measured the circumference of Earth. BACKGROUND NOTES 4.

2. Describe the geocentric models of the solar system developed by Aristotle and Ptolemy. TEXTBOOK PAGES 57–61; VIDEO PROGRAM.

3. Describe the motions of the naked-eye planets as observed from Earth and define the term retrograde. TEXTBOOK PAGES 59–60; VIDEO PROGRAM.

4. Explain the Copernican model and hypothesis. TEXTBOOK PAGES 61–64; VIDEO PROGRAM.

5. Compare and contrast the Ptolemaic (geocentric) model of the solar system with the Copernican (heliocentric) model. TEXTBOOK PAGES 59–64.

6. Identify the various instruments used by Tycho Brahe to make his observations and describe the various conclusions he made based on his observations. TEXTBOOK PAGES 64–67; VIDEO PROGRAM.

7. Explain Kepler's three laws of planetary motion. TEXTBOOK PAGES 67–70; VIDEO PROGRAM.

8. Describe the telescopic observations of the solar system made by Galileo. TEXTBOOK PAGES 70–74; VIDEO PROGRAM.

9. Distinguish between hypothesis, model, theory, and law. TEXTBOOK PAGE 63.

Viewing Notes

This video program surveys the roots of the modern science of astronomy in the contributions of Aristotle, Ptolemy, Copernicus, Galileo, Tycho Brahe, and Kepler.

"The Origin of Modern Astronomy" begins by examining the ideas of Aristotle, who was one of the first Greeks to apply geometric thinking to explain the motions he observed in the sky. Aristotle's ideas were perpetuated and refined by Ptolemy, who developed a more comprehensive model of the universe. The program explains and illustrates how the rather complex Ptolemaic model accounted for the motions of the planets.

Most of the remainder of the program covers the 99 years, from 1543 to 1642, in which the true foundation of modern astronomy was built from the contributions of Copernicus, Tycho Brahe, Kepler, and Galileo.

The video program first features Nicholaus Copernicus. Several astronomers explain the features of his heliocentric model of the universe and compare it to the earlier Ptolemaic model. Computer animation clearly illustrates how the Copernican model explains the retrograde motion of the planets.

Although the Copernican model was the first that correctly placed the sun at the center of the universe (our solar system), the idea was not widely accepted until Galileo's observations showed that Venus orbited the sun and that objects (moons) orbited Jupiter. These observations clearly suggested that Earth orbited the sun. The video program describes his telescopic observations of the moon and planets and explains how they supported the Copernican model.

Both Copernicus and Galileo experienced personal difficulties with the Roman Catholic Church because of their models and observations. The video program describes how their work conflicted with church dogma.

The video program concludes with a description of the work of Tycho Brahe and Johannes Kepler. Tycho's theories, although in opposition to the Copernican model, were based on meticulous observations made with precise instruments. After Tycho's death, Kepler studied Tycho's observations and developed his three laws of planetary motion, which were the first to predict accurately the motions of the planets.

The following timeline may be helpful in tracing the beginnings of modern astronomy.

c. 580 B.C.	Pythagoras born
500 B.C.	Pythagoras dies
384 B.C.	Aristotle born

322 B.C.	Aristotle dies
200 B.C.	Eratosthenes measures circumference of Earth
A.D. 140	Ptolemy develops model of universe
1473	Nicolaus Copernicus born
1507	Copernicus publishes theory that the sun is the center of the universe
1543	Copernicus dies
1546	Tycho Brahe born
1564	Galileo Galilei born
1571	Johannes Kepler born
1573	Tycho Brahe publishes *De Stella Nova*
1600	Kepler starts to work with Tycho Brahe
1601	Tycho Brahe dies
1609	Kepler publishes *Astronomia Nova*
1610	Galileo publishes *Sidereus Nuncius* describing his early telescopic observations
1633	Galileo sentenced to life imprisonment (house arrest)
1642	Galileo dies

As you watch the video program, consider the following questions:

1. How did the Greek approach to science differ from that of other ancient civilizations?

2. What basic questions were the early astronomers trying to answer?

3. What were the components of the Ptolemaic model of the solar system and how did the model account for the backward, or retrograde, motion of the planets?

4. What were the main features of the Copernican model of the solar system and how did it account for the retrograde motion of the planet Mars?

5. What were the advantages and limitations of the Copernican model?

6. How did Galieo's discovery of satellites around Jupiter and the phases of Venus support the heliocentric model of the solar system?

7. Who was Tycho Brahe and what contributions did he make toward our understanding of the solar system?

8. What are Kepler's three laws of planetary motion?

Review Activities

Matching

Match the terms listed below with the definitions that follow. Check your answers with the Answer Key and review any terms you missed.

_f_ 1. parallax   _a_ 8. heliocentric universe

_c_ 2. geocentric universe   _g_ 9. hypothesis

___ 3. uniform circular motion   _e_ 10. model

_h_ 4 retrograde motion   _l_ 11. theory

_b_ 5. epicycle   _j_ 12. natural law

_m_ 6. deferent   _d_ 13. ellipse

_k_ 7. equant

a. a model of the universe with the sun at the center, such as the Copernican universe

b. in the Ptolemaic theory, a small circle in which a planet moves

c. a model of the universe with Earth at the center

d. a closed curve around two points called foci, such that the total distance from one focus to the curve and back to the other focus remains constant

e. a description of a natural phenomenon

f. the apparent change in the observed position of an object due to a change in the location of the observer

g. a conjecture, subject to testing, to explain certain facts

h. the apparent backward (westward) motion of planets as seen against the background stars

i. the belief that objects in the sky move uniformly along perfectly circular orbits

j. a theory that is almost universally accepted as true

k. in the Ptolemaic theory, the point off center in the deferent from which the center of the epicycle appears to move uniformly

l. a system of assumptions and principles applicable to a wide range of phenomena that have been repeatedly verified

m. in the Ptolemaic theory, the large circle around Earth along which the center of the epicycle moved

Completion

Fill each blank with the most appropriate term from the list for that paragraph. A term may be used once, more than once, or not at all. Check your answers with the Answer Key and review when necessary.

1. Ptolemy developed what is commonly called the _geocentric_ or _Earth_ -centered system. In this model of the solar system, Earth remained _stationary_, and the planets, sun, and moon moved in _circular_ orbits around Earth. Because the Greeks believed the heavens were perfect, Ptolemy's model required the planets to move at _constant_ speed. This movement of the planets is referred to as _uniform circular_ motion.

    However, the planets, at times, do not always appear to move forward. Occasionally, the planets appear to move backward or _retrograde_ against the stars. To explain this perceived backward motion, Ptolemy placed the planets on small circles called _epicycles_. In order to maintain the notion of _constant_ speed, Ptolemy placed these small circles on larger circles called _deferents_.

    Because motions of the planets still did not match the observations satisfactorily. Ptolemy adjusted his model by moving Earth _off center_. Ptolemy's model was very complicated. Many circles upon circles were required to explain why the planets were observed to move at _constant_ speed. However, this desired motion could be viewed only from a point called the _equant_, a point that was not on Earth.

circular	heliocentric
constant	off center
deferents	retrograde
Earth	stationary
epicycles	sun
equant	uniform circular
geocentric	

2. Johannes Kepler used the observational data of _Tyco Brahe_ to devise _Three_ laws that described the motions of the planets. These laws expanded upon the _Heliocentric_ or _Sun_ centered model of Copernicus. Kepler's laws were the first to connect observation with mathematical relationships. In other words, they described the solar system as it truly is.

Kepler's first law states that the orbits of the planets are _ellipses_ with the _Sun_ as one of the foci. The second law states that a line from the planet to the sun will sweep out _Equal_ areas in _equal_ times. This law means that when the planet is closest to the sun, or at _Perihelion_, the planet travels _fastest_ in its orbit. When the planet is farthest from the sun, the planet travels _Slowest_ in its orbit. The third law states that the _Square_ of the orbital period of the planet is proportional to the _cube_ of the average distance of the planet from the sun.

It is important to remember that Kepler's laws are _empirical_. In other words, Kepler determined them solely from _observations_. They were not _derived_ from a more fundamental law.

Copernicus	geocentric
cube	heliocentric ✓
derived	observations
Earth	perihelion
ellipses	slowest
empirical	square
equal	sun
fastest	three ✓
five	Tycho Brahe ✓

Self-Test

Select the one best answer for each question. Check your answer with the Answer Key and review when necessary.

1. Eratosthenes based his measurement of the circumference of Earth on the observed position of the sun at the same time and on the same day in the cities of Alexandria and Syene. He observed that the sun at Syene was

   a. on the horizon and at the zenith at Alexandria.

   b. at the zenith and on the horizon at Alexandria.

   c. at the zenith and 7 degrees south of (below) the zenith at Alexandria.

   d. in the same position in the sky as it was in Alexandria.

2. The unit of distance Eratosthenes used to calculate the circumference of Earth was the

   a. stadium.

   b. degree.

   c. kilometer.

   d. number of degrees in a complete circle.

3. Aristotle concluded that Earth remained stationary because he was unable to observe parallax in the stars. Parallax is the

   a. motion of the celestial sphere around the stationary Earth.

   b. relationship between the velocity of a planet and its distance from Earth.

   c. change in the apparent position of an object due to a change in the location of the observer.

   d. change in the position of a star due to the season of the year during which the observation is made.

4. Ptolemy assumed that the planets moved through the perfect heavens in perfect motion. This perfect motion is called

   a. uniform circular motion.

   b. uniform accelerated motion.

   c. motion without parallax.

   d. retrograde motion.

5. In order to account for some of the observed peculiar motions of the planets, Ptolemy's model required that the planets move in

   a. small circles called deferents.

   b. straight lines at constant velocity.

   c. small circles called epicycles.

   d. the equant.

6. As a result of Earth's rotation, the stars appear to rise in the east and set in the west. If we observe the motion of the planets against the background stars, the planets usually appear to move from

   a. east to west.

   b. west to east.

   c. north to south.

   d. north to west.

7. Planets are observed to move, for a time, in retrograde motion. This motion describes the

   a. easterly drift of the planets.

   b. westerly drift of the stars.

   c. backward or easterly motion of the stars.

   d. backward or westerly motion of the planets.

8. In *De Revolutionibus*, Copernicus proposed that the sun was located at

   a. one of the foci of an ellipse.

   b. the center of the universe.

   c. directly above Earth.

   d. the equant.

9. The movements of the planets in the heliocentric system of Copernicus required

   a. elliptical orbits.

   b. parabolic orbits.

   c. straight-line motion.

   d. uniform circular motion.

10. Epicycles were used to describe the motions of the planets by
   a. Copernicus.
   b. Ptolemy.
   c. both Copernicus and Ptolemy.
   d. astronomers before the time of Ptolemy.

11. The Copernican model of the solar system eventually gained acceptance over the Ptolemaic model. The probable main reason for this acceptance was that the
   a. accuracy of the predictions were better than the Ptolemaic model.
   b. sun was placed at one of the foci of the planetary orbits.
   c. sun was placed at the center of the solar system.
   d. whole idea of the heliocentric system was pleasing to the eye and intellect.

12. The success of Tycho Brahe is attributed to his ability to
   a. make accurate observations with large and well-constructed instruments.
   b. construct theoretical models of physical phenomena.
   c. receive financial support from the king for his astronomical endeavors.
   d. make excellent observations with a telescope.

13. According to the Tychonic model of the solar system,
   a. Earth was the center of the universe and the planets orbited Earth.
   b. Earth was stationary, the sun and moon orbited Earth, and the planets orbited the sun.
   c. the sun was the center of the universe and the planets orbited the sun in elliptical paths.
   d. the sun was the center of the universe and the planets orbited the sun in circular paths.

14. According to Kepler's first law of planetary motion, the orbits of the planets are

    a. circles, with Earth located at the center.

    b. ellipses, with the sun located at one of the foci.

    c. ellipses, with Earth located at one of the foci.

    d. are circles, with the sun located at the center.

15. A planet moves fastest in its orbit when it is

    a. closest to Earth.

    b. farthest from the sun.

    c. at one of the foci.

    d. closest to the sun.

16. Galileo's most convincing observation that proved Venus orbited the sun was the discovery of

    a. moons in orbit around Jupiter.

    b. the highly reflective atmosphere of Venus.

    c. the phase cycle of Venus.

    d. evidence for the rotation of the planet.

17. Galileo observed and recorded

    a. sunspots, satellites around Jupiter, phases of Venus, lunar craters and mountains.

    b. dark lines on Venus, phases of Jupiter, lunar craters and mountains.

    c. satellites around Jupiter, phases of Venus, sunspots, lunar craters and mountains.

    d. sunspots, mountains on Venus, lunar craters, satellites around Jupiter.

18. A single conjecture or assertion that must be tested best describes a

    a. law.

    b. theory.

    c. hypothesis.

    d. model.

19. A model can best be described as a

   a. description of a natural phenomenon.

   b. representation of what actually exists in nature.

   c. correct physical representation of a conceptual idea.

   d. statement of fact that has been tested and shown to be correct.

Using What You've Learned

According to your own interests or your instructor's assignment, complete one or more of the following activities.

1. Follow the instructions below to construct an ellipse, the basic geometrical shape for the orbits of the planets, comets, and asteroids.

   a. Cut a piece of strong string about 10 inches in length and tie the ends together.

   b. Place a sheet of paper on a board and push two thumbtacks, about 2 inches apart, into the board. These locations are called the foci.

   c. Place the loop of string around the outside of the two thumbtacks.

   d. Use a pencil to stretch the piece of string into the shape of a triangle. The pencil is one vertex and the two thumbtacks are the other two vertices. The string forms the sides of the triangle.

   e. Move the pencil, keeping the string taut, and allow the string to guide the pencil to draw an ellipse. Keep the pencil in contact with the paper at all times and make sure the pencil and thumbtacks remain the vertices of a triangle and the string remains as the sides of the triangle. Continue in this manner until you have completed an ellipse. If part of the ellipse falls off the paper, shorten the string. Refer to Figure 4–16(a) on page 69 in the *Horizons* textbook.

   f. After drawing the ellipse, measure the major axis (the largest diameter through the two foci = 2a) and the distance between the two foci (2c).

   g. Calculate the eccentricity of the ellipse:

   $$e = c/a$$

2. In the following diagram of a planetary orbit, locate and label:

   a. one possible position of the sun (focus)

   b. the major axis

   c. the position where the planet would have its slowest speed.

   d. perihelion

   e. the position where the planet would have its greatest speed

   f. aphelion

PLANETARY ORBIT

3. Develop a diagram that shows the occurrence of the phases of Venus. If Venus is currently visible, use a pair of binoculars to observe the planet and determine where on your diagram Venus must be in order for you to observe its current phase.

4. Where on Earth:

   a. do the stars rise in the west?

   b. is the south celestial pole at the south point on the horizon?

   c. is the sun at the north point on the horizon at midnight on the day of the summer solstice?

   d. is the zenith directly overhead?

   e. does the sun pass through the zenith on the day of the summer solstice?

   f. is the observer's latitude equal to the altitude of the north celestial pole (assume the observer is in the northern hemisphere)?

   g. does the sun rise once a year on the autumnal equinox?

   h. are all the stars above the horizon half of the time?

Answer Key

Matching

1.	f	8.	a
2.	c	9.	g
3.	i	10.	e
4.	h	11.	l
5.	b	12.	j
6.	m	13.	d
7.	k		

Completion

1. geocentric, Earth, stationary, circular, constant, uniform circular, retrograde, epicycles, constant, deferents

    off center, constant, equant

2. Tycho Brahe, three, heliocentric, sun

    ellipses, sun, equal, equal, perihelion, fastest, slowest, square, cube

    empirical, observations, derived

Self-Test

(Page numbers refer to the textbook.)

1. c (Objective 1; Background Notes 4)
2. a (Objective 1; Background Notes 4)
3. c (Objective 2; page 58)
4. a (Objective 2; page 59)
5. c (Objective 2; pages 59–60; video program)
6. b (Objective 3; page 59)
7. d (Objective 3; page 59; video program)
8. b (Objective 4; page 61)
9. d (Objective 4; page 59; video program)
10. c (Objective 5; pages 60, 62; video program)
11. d (Objective 5; page 62; video program)

12. a (Objective 6; page 65; video program)

13. b (Objective 6; page 65; video program)

14. b (Objective 7; page 68)

15. d (Objective 7; page 69)

16. c (Objective 8; page 72; video program)

17. a (Objective 8; pages 71–72; video program)

18. c (Objective 9; page 63)

19. a (Objective 9; page 63)

5

Newton, Einstein, and Gravity

Assignments

For the most effective study of this lesson, we suggest that you complete the assignments in the sequence listed below.

Before Viewing the Video Program

- Read the Overview and Learning Objectives for this lesson. Use the Learning Objectives to guide your reading, viewing, and thinking.

- Read Chapter 4, "The Origin of Modern Astronomy," sections 4–5 and 4–6, pages 70–77, in the textbook.

- Read Background Notes 5A, "The Law of Universal Gravitation," and 5B, "Special and General Relativity," following Lesson 6 in this guide.

- Read the Viewing Notes in this lesson.

View "Newton, Einstein, and Gravity" Video Program

After Viewing the Video Program

- Briefly note your answers to questions listed at the end of the Viewing Notes.

- Review all reading assignments for this lesson, especially the textbook Chapter 4 summary on pages 77–78, the Background Notes 5A and 5B, and the Viewing Notes in this lesson.

- Write a brief answer to review question 10 at the end of the textbook Chapter 4 to be certain you understand the text material.

- Complete the Review Activities in this guide to reinforce your understanding of important terms and concepts. Check your answers with the Answer Key at the end of this lesson and review when necessary.

- Take the Self-Test in this guide to measure your achievement of the Objectives. Check your answers with the Answer Key and review when necessary.

- Complete the Using What You've Learned activities and any other activities and projects assigned by your instructor.

Overview

Johannes Kepler developed a set of mathematical rules that described how planets move around the sun. Galileo Galilei developed a set of mathematical rules that described how objects fall near the surface of Earth. Galileo's law of falling bodies was quite contrary to the ideas of motion held by Aristotle. Kepler's and Galileo's rules established the concept that mathematical relationships could be used to describe the behavior of the universe. However, these relationships were limited because they could describe only nearby events. They could not explain the causes of that motion. That next step in our understanding of the universe was made by Isaac Newton.

In his published work, the *Principia*, Newton put forth an explanation for the causes of motion. He introduced the ideas of force, mass, and acceleration. He redefined the nature of objects in motion and at rest. Newton spent nearly two decades developing his laws of motion, making fundamental discoveries in optics and mathematics and deriving a law that describes one of the most, if not the most important force in the universe: gravity. It is through understanding and applying gravity that scientists have been able to calculate the return of comets, predict the presence of unseen planets, land humans on the moon, and launch spacecraft that escape from the solar system.

Along with our ability to describe the universe with mathematical relationships, we also developed the technology to observe the universe in greater detail. It is the result of these observations that brought the laws developed by Isaac Newton into question. For instance, Newton's law of universal gravitation could not satisfactorily explain some phenomena, and modern science found inconsistencies between Newton's laws of motion and the behavior of small atomic particles traveling at high speeds. It was the general and special theories of relativity of Albert Einstein that explained the new observations and resolved the inconsistencies. Because Einstein's theories required few fundamental postulates, the theories are aesthetically pleasing to scientists. However simple these postulates seem to be, they are very powerful and provide us with insight into areas beyond the realm of our personal experiences and intuition. The logical, observational, and experimental evidence in support of these theories has elevated them to the status of laws.

This entire lesson is devoted to developing an understanding of the classical (Newton) and modern (Einstein) view of the universe. Newton's laws of motion and Einstein's theories of relativity describe the nature of the universe, can be expressed mathematically, and are supported by experimental evidence. With these laws as a foundation, we can now explore everything from the nature of planets and their orbits about the sun to the nature of stars and their evolution. We can even begin to develop a description and understanding of the origin, evolution, and destiny of the universe.

Learning Objectives

When you have completed all assignments for this lesson, you should be able to:

1. Describe Newton's three laws of motion. TEXTBOOK PAGE 76.

2. State Newton's law of universal gravitation and describe how it explains orbital motion. TEXTBOOK PAGES 76–77; BACKGROUND NOTES 5A.

3. Identify the various astronomical observations and calculations performed in the eighteenth and nineteenth centuries that initially supported the validity of Newton's law of universal gravitation. BACKGROUND NOTES 5A; VIDEO PROGRAM.

4. State the basic postulates of Einstein's special theory of relativity. BACKGROUND NOTES 5B; VIDEO PROGRAM.

5. Explain and describe the following consequences of special relativity: time dilation and length contraction. BACKGROUND NOTES 5B; VIDEO PROGRAM.

6. Explain the relationship $E = mc^2$. BACKGROUND NOTES 5B; video program.

7. State the basic postulates of Einstein's general theory of relativity. BACKGROUND NOTES 5B; VIDEO PROGRAM.

8. Explain what is meant by the curvature of space-time and describe the following consequences of general relativity: precession of orbits (Mercury), bending of starlight, time dilation and gravitational red shift, and gravitational radiation (gravity waves). BACKGROUND NOTES 5B; VIDEO PROGRAM.

Viewing Notes

"Newton, Einstein, and Gravity" is about the quest of science to understand the nature of gravity and motion. The first part of the video program traces how early scientists, including Aristotle, Galileo, and Kepler, tried to explain gravity and motion.

Most of the video program focuses on the contributions of two scientists: Isaac Newton and Albert Einstein. Newton—in his universal law of gravity and three laws of motion—was the first person to explain successfully such phenomena as elliptical orbits and the motions of all objects. Film of astronauts in an orbiting space shuttle clearly illustrate Newton's three laws of motion and their universality. Computer graphics help to explain Newton's other major contribution: the law of universal gravitation and how it describes orbits.

Although Newton's laws worked for the universe as it was known in his time, they did not explain phenomena that scientists were able to observe around the beginning of the twentieth century, such as atoms and electrons moving at near the speed of light. It was Einstein who developed theories to explain what Newton's laws could not explain.

In the video program, several physicists and astronomers explain Einstein's special and general theories of relativity. Computer graphics illustrate key aspects of each theory, including the universality of the speed of light, time dilation, the twin paradox, equivalence of acceleration and gravity, and the curvature of space-time.

The video program concludes with two striking "proofs" of Einstein's theories: how the general theory of relativity explains the precession of Mercury's orbit and how the sun's gravity bends light from stars behind it.

As you watch the video program, consider the following questions:

1. What is the difference between a theory and a law?

2. What are Newton's three laws of motion? What is Newton's law of universal gravitation?

3. What is the principle of relativity? What did Einstein say about the speed of light as measured by different observers?

4. What is the twin paradox?

5. How can muons reach the surface of Earth?

6. What do we mean when we say that gravity and acceleration are equivalent?

7. What do we mean by space-time? What do we mean when we say that mass curves space-time around it?

8. How was a solar eclipse used to observe the bending of starlight around the sun?

Review Activities

Matching

Match the terms listed below with the definitions that follow. Check your answers with the Answer Key and review any terms you missed.

__d__ 1. mass

__k__ 2. weight

__b__ 3. inverse square law

__h__ 4. *Principia*

__H__ 5. Newton's first law of motion

__o__ 6. Newton's second law of motion

__g__ 7. Newton's third law of motion

__m__ 8. law of universal gravitation

__i__ 9. space-time

__a__ 10. time dilation

__L__ 11. $E = mc^2$

__e__ 12. length contraction

__j__ 13. gravitational red shift

__C__ 14. gravity waves

__f__ 15. curvature of space-time

__P__ 16. principle of equivalence

a. the slowing of time as measured by one observer in motion relative to another

b. a rule that the strength of an effect (such as gravity) decreases in proportion as the distance squared increases

c. ripples in space-time that may travel outward from sources of gravity

d. a measure of the amount of matter in an object

e. the shortening of length in the direction of motion as measured by an observer considered to be at rest

f. the warping of space in the presence of mass

g. when one body exerts a force on another, the second body exerts an equal but opposite force on the first body

h. a body continues at rest or in uniform motion in a straight line unless acted upon by some force

i. the four dimensions in which all things (matter and energy) exist; one dimension for time and the other three for possible directions (motions) in space

j. the lengthening of the wavelength of a photon (decrease in frequency) due to its escape from a gravitational field

k. the force of gravity acting on an object on or near a planet or moon; it varies from place to place and planet to planet

l. relationship between mass and energy

m. the force between two objects is determined by their masses and the inverse square of the distance between them

n. publication by Isaac Newton containing his work on laws of motion and universal gravitation

o. a body's change of motion is proportional to the force acting on it and is in the direction of the force

p. basic postulate of general relativity describing the relationship between a gravitational force and a constant acceleration

Completion

Fill each blank with the most appropriate term from the list for that paragraph. A term may be used once, more than once, or not at all. Check your answers with the Answer Key and review when necessary. If a question requires two or more answers in succession, they may be in any order, unless the question indicates otherwise.

1. Isaac Newton developed, among other things, three laws to explain motion and one law to describe gravity. His first law describes the nature of motion of an object. It states that if an object is at rest it will remain at _rest_. If the object is in motion, it will _continue_ in motion at a _constant_ speed and in a constant _direction_ until it is acted upon by some _force_. According to Newton's second law, the _change_ in a body's motion is _proportion_ to the force acting on the body. Furthermore, that _change_ in the motion is in the _same_ direction as the force acting on the body. If a force is applied to an object, the object exerts a _force_, _equal_ in size to the original _force_ but _opposite_ in direction, on the first body. This _action reaction_ law is known as Newton's third law of motion. Newton explained the nature of gravity in his law of _universal_ gravitation. The force of gravity is _proportional_ to the _masses_ of the objects and _inversely_ proportional to the _square_ of the distance between them.

action-reaction	masses
change	opposite
constant	proportional
continue	rest
conversely	same
direction	special
equal	square
force	universal
inversely	variable

2. According to Einstein's special theory of relativity, the laws of physics
   are the _Same_ for all observers, and the _Speed_ of
   light is also the same for all observers regardless of their
   relative motions. These two fundamental postulates require us to
   reevaluate the nature of _Space_ and _time_. The effects of
   special relativity are important when objects travel at speeds near that
   of _light_. The phenomenon of time _Dilation_ results in the
   Slowing of clocks in a _Moving_ spacecraft when compared
   with similar clocks that are considered to be at _rest_. An
   example of this phenomenon is found in the motion of the subatomic
   particle called the _Muon_. Its lifetime is so _brief_, it should
   not be able to reach the surface of Earth when produced in the upper
   atmosphere. However, because of _time Dilation_ the clock of the
   Muon runs _Slower_ and the particle lives long enough to
   reach the ground. Another important result of special relativity
   demonstrates the relationship between mass and _energy_. In the
   famous equation $E = mc^2$, the "E" stands for _energy_, the "m"
   stands for _mass_, and the "c" stands for the speed of
   light. Once this equation was understood, scientists could
   explain why the _Stars_ shine for as long as they do.

brief	rest
dilation	same
$E = mc^2$	slower
energy	slowing
faster	space
light	speed
long	stars
mass	time
moving	time contraction
muon	time dilation

3. Einstein's general theory of relativity is based on the postulate that equates _Gravity_ and _Acceleration_. This statement is called the _Principle of Equivalence_. It states that it is _impossible_ to distinguish between a constant _Acceleration_ and being at rest in a _Gravitational_ field. General relativity is our best way to describe gravity. It describes how gravity produces a _curvature_ of space-time and that space-time consists of _four_ dimensions. General relativity also explains why the orbit of Mercury _precesse_ around the sun, how gravity can _bend_ starlight, and how time _slows_ in a gravitational field. The effect gravity has on time is known as a gravitational _red shift_. Furthermore, general relativity makes a prediction about the existence of _gravity_ waves. These waves are ripples in _space-time_ and travel outward, at the speed of _light_, from the explosion of a star or from two stars in orbit about each other. Although these waves have yet to be detected directly, their existence is indirectly verified by the observation of two _neutron_ stars in orbit about each other. The orbit slowly _decays_ by the amount predicted by general relativity.

acceleration	light
bend	neutron
curvature	precesses
decaying	principle of equivalence
decays	red shift
four	six
gravitational	slows
gravity	space-time
impossible	two
increases	

Self-Test

Select the one best answer for each question. Check your answers with the Answer Key and review when necessary.

1. If an object is moving and it remains undisturbed by outside forces, it will

   a. stop. *WRONG*

   b. continue to move in a circular path.

   ✓ c. continue to move in a straight line with a uniform velocity.

   d. stop and begin to move in the opposite direction.

2. If you double the force acting on an object, assuming the mass remains constant, the acceleration of that object would be

   a. doubled.

   b. halved.

   c. four times greater.

   d. one-fourth as great.

3. According to Newton's law of universal gravitation, if the distance between two objects increases, the force of gravity between them will

    a. increase.

    b. decrease by a factor equal to the square of the distance between the objects.

    c. increase by a factor equal to the square of the distance between the objects.

    d. remain unchanged.

4. A satellite launched into orbit around Earth, far above the effects of Earth's atmosphere, will

    a. eventually escape Earth's gravity and orbit the moon.

    b. eventually escape Earth's gravity and go into orbit around the sun.

    c. continue to orbit Earth.

    d. remain fixed in space as Earth rotates beneath the satellite.

5. The discovery of Neptune was based on the observed gravitational perturbations of the planet

    a. Mars.

    b. Jupiter.

    c. Saturn.

    d. Uranus.

6. One of the earliest uses of Newton's law of universal gravitation was the calculation of the orbit of a comet by

    a. Edmund Halley.

    b. Isaac Newton.

    c. John C. Adams.

    d. William Herschel.

7. The speed of light
    a. varies with distance from the observer.
    b. remains constant for all observers.
    c. decreases with time.
    d. increases when two objects are observed to be approaching each other.

8. Although observers may move at uniform speeds relative to each other, they will always find that
    a. time is absolute.
    b. mass cannot transform into energy.
    c. the laws of physics depend on the relative motions of observers.
    d. the laws of physics are the same for all observers regardless of their relative uniform motion.

9. An example of time dilation can be observed by the fact that
    a. muons survive the journey from the upper atmosphere to the surface of Earth.
    b. mass can be transformed into energy.
    c. Halley's comet orbits the sun once every 76 years.
    d. the Michelson-Morely experiment did not detect the ether.

10. According to the special theory of relativity, *moving* clocks
    a. run more slowly at all times.
    b. run more slowly when the observer is moving at extremely high velocities.
    c. run more slowly when compared to other clocks that are considered to be at rest.
    d. never run more slowly . The effect is an illusion. Time runs at the same rate for everyone.

11. The "E" in the relationship $E = mc^2$ refers to the energy
    a. produced if all the mass "m" is converted to energy.
    b. associated with the object's motion.
    c. absorbed in an atomic explosion.
    d. produced in a nuclear power plant.

12. One of the most immediate applications of the equation $E = mc^2$ to astronomical phenomena was to explain the source of energy in

    a. orbiting planets.

    b. the interior of stars.

    c. muons and electrons.

    d. black holes and neutron stars.

13. It is impossible to distinguish between a uniform acceleration and being at rest in a gravitational field is a statement of the

    a. time dilation effect.

    b. principle of equivalence.

    c. length contraction phenomenon.

    d. curvature of space-time.

14. Space-time is said to consist of four dimensions. These four are

    a. two time dimensions and two space dimensions.

    b. three time dimensions and one space dimension.

    c. three space dimensions and one time dimension.

    d. four space dimensions.

15. In the absence of mass, space-time is said to be flat. In the presence of mass, space-time is said to be

    a. smooth.

    b. length contracted.

    c. confused with accelerations.

    d. curved.

16. According to Newton's law of universal gravitation, a beam of light is considered to have no mass and therefore would be unaffected by gravity. General relativity, however, predicts that light will be affected by gravity. This prediction is demonstrated by the fact that light

    a. travels only in straight lines.

    b. is observed to have a constant speed.

    c. creates length contraction effects.

    d. bends around massive objects.

17. Hypothetical ripples in space-time produced by supernova events or orbiting massive stars, which propagate outward at the speed of light, are referred to as

    a. gravity waves.

    b. neutron stars.

    c. black holes.

    d. gravitational red shifts.

Using What You've Learned

Provide short answers to the following questions about special and general relativity.

1. Why was the Michelson-Morely experiment considered a failure?

2. How did G. F. Fitzgerald explain the failure of the Michelson-Morely experiment?

3. What is the fundamental postulate concerning the speed of light in Einstein's special theory of relativity?

4. What is meant by space-time?

5. State the principle of equivalence in your own words.

6. Take a ride on a roller coaster, such as the "Star Tours" ride at Disneyland or Disney World, or any "multimedia" ride that simulates motion with visual aids. List the various times during the ride when accelerations simulate gravity and when gravity simulates accelerations.

7. What are the important differences between special and general relativity?

8. Pay close attention to your activities for a period of several hours. Make a list of ten significant activities you did during this time. Explain how Newton's three laws of motion related to your activities.

9. Watch several cartoon programs such as the *Road Runner* and *Tom and Jerry*. Identify the various scenes that violate Newton's laws of motion. Explain what would happen to the cartoon characters if the laws were allowed to occur correctly. (Ignore injuries and possible death to the cartoon characters.)

Answer Key

Matching

1.	d	9.	i
2.	k	10.	a
3.	b	11.	l
4.	n	12.	e
5.	h	13.	j
6.	o	14.	c
7.	g	15.	f
8.	m	16.	p

Completion

1. rest, continue, constant, direction, force, change, proportional, change, same, force, equal, force, opposite, action-reaction, universal, proportional, masses, inversely, square

2. same, speed, light, space, time, light, dilation, slowing, moving, rest, muon, brief, time dilation, muon, slower, energy, energy, mass, light, stars

3. gravity, acceleration, principle of equivalence, impossible, acceleration, gravitational, curvature, four, precesses, bend, slows, red shift, gravity, space-time, light, neutron, decays

Self-Test

(Page numbers refer to the textbook.)

1. a   (Objective 1; page 76)

2. a   (Objective 1; page 76)

3. b   (Objective 2; page 76)

4. c   (Objective 2; pages 76–77)

5. d   (Objective 3; Background Notes 5A; video program)

6. a   (Objective 3; Background Notes 5A; video program)

7. b   (Objective 4; Background Notes 5B; video program)

8. d (Objective 4; Background Notes 5B; video program)

9. a (Objective 5; Background Notes 5B; video program)

10. c (Objective 5; Background Notes 5B; video program)

11. a (Objective 6; Background Notes 5B; video program)

12. b (Objective 6; Background Notes 5B; video program)

13. b (Objective 7; Background Notes 5B; video program)

14. c (Objective 7; Background Notes 5B; video program)

15. d (Objective 8; Background Notes 5B; video program)

16. d (Objective 8; Background Notes 5B; video program)

17. a (Objective 8; Background Notes 5B; video program)

L E S S O N
6

The Tools of Astronomy

Assignments

For the most effective study of this lesson, we suggest that you complete the assignments in the sequence listed below.

Before Viewing the Video Program

- Read the Overview and Learning Objectives for this lesson. Use the Learning Objectives to guide your reading, viewing, and thinking.
- Read Chapter 5, "Astronomical Tools," pages 80–108, in the textbook.
- Read the Viewing Notes in this lesson.

View "The Tools of Astronomy" Video Program

After Viewing the Video Program

- Briefly note your answers to questions listed at the end of the Viewing Notes.
- Review all reading assignments for this lesson, especially the textbook Chapter 5 summary on page 106 and the Viewing Notes in this lesson.
- Write brief answers to all the review questions at the end of textbook Chapter 5 to be certain you understand the text material.
- Complete the Review Activities in this guide to reinforce your understanding of important terms and concepts. Check your answers with the Answer Key at the end of this lesson and review when necessary.

Lesson 6 / The Tools of Astronomy   **75**

- Take the Self-Test in this guide to measure your achievement of the Objectives. Check your answers with the Answer Key and review when necessary.

- Complete any other activities and projects assigned by your instructor.

Overview

The course thus far has been almost entirely about the history of astronomy—how humanity's quest to understand the universe was transformed from an almost mystical preoccupation in ancient times to the first true science during the Renaissance. Even when discussing natural phenomena such as eclipses and seasons, the course has concentrated on naked-eye phenomena, which have been thoroughly observed and studied by our ancestors.

It should be clear that the turning point in astronomy was Galileo's use of the telescope in the early seventeenth century. Although his instrument was crude compared to even a toy telescope today, it still revealed the cosmos with about ten times more clarity than the human eye alone could discern. Thus, almost everything Galileo observed was a major discovery that often contradicted some cherished ancient belief. Although professional telescopes today are vastly larger and more powerful than Galileo's, our observational capabilities have grown slowly and in small increments over the centuries. For centuries, Galileo's humble spyglass remained the single greatest technological leap in astronomical resolving power, until the successful repair of the Hubble Space Telescope in 1993 achieved a similarly great leap.

Today, the telescope is still the single most important tool available to astronomers who, unlike other scientists, can look at, but not touch, the objects they study. It is actually the light from distant stars that astronomers "touch." It is the light that has traveled through space to Earth that astronomers have in their "possession" and that they concentrate and analyze with instruments. Observational astronomy is a quest for light. This lesson shows that every improvement in astronomical technology has been the result of an attempt to gather more light, to gather it over a wider range of wavelengths, or to focus it in a way that reveals greater detail. As you will see throughout the remainder of the course, every improvement in technology has led to a corresponding improvement in our understanding of the universe.

Learning Objectives

When you have completed all assignments for this lesson, you should be able to:

1. Explain the characteristics of the electromagnetic spectrum and their effects on astronomical equipment and observations. TEXTBOOK PAGES 80–83; VIDEO PROGRAM.

2. Describe the operation of optical telescopes, including the relative advantages and disadvantages of refracting versus reflecting designs, different focus locations, and mountings. TEXTBOOK PAGES 83–90.

3. Describe the factors that determine the light-gathering, magnifying, and resolving powers of a telescope, including the effects of Earth's atmosphere. TEXTBOOK PAGES 90–93; VIDEO PROGRAM.

4. Describe the instrumentation used to accomplish photography, photometry, and spectroscopy with a telescope. TEXTBOOK PAGES 93–96; VIDEO PROGRAM.

5. Describe the operation, limitations, and advantages of radio telescopes, including the principle of interferometry. TEXTBOOK PAGES 96–100; VIDEO PROGRAM.

6. Describe space-based telescopes designed for both visible and nonvisible astronomy. TEXTBOOK PAGES 100–106; VIDEO PROGRAM.

Viewing Notes

"The Tools of Astronomy" begins with an overview of the electromagnetic spectrum, from gamma rays to radio waves, and the types of information provided by individual bands. Astronomers also describe how Earth's atmosphere creates problems for them in making observations.

Most of the remainder of the video program focuses on the major types of telescopes used by astronomers, while showing you some of the greatest observatories in the world and the environments in which astronomers carry out their explorations.

The first telescope visited in the video program is the refracting telescope at Lick Observatory in northern California. This telescope, one of the oldest large telescopes in the world, was the first to be placed on a mountaintop to avoid some of the distortion created by Earth's atmosphere. The telescope's lens, with a diameter of 36 inches, is near the upper limit of practical size for distortion-free images.

The video program next shows some of the most important and largest reflecting telescopes in use today, at Kitt Peak National Observatory in northern Arizona and at the Keck Observatory in Hawaii. Staff astronomers describe the features of the telescopes and explain how they operate. Astronomers also discuss how charge-coupled devices (CCDs), which feed faint images directly into computers, and adaptive optics enhance the images received by large telescopes.

Radio telescopes are featured in the next segment, which describes how these telescopes collect radio waves—the only other wavelengths, besides visible light, that easily penetrate Earth's atmosphere. The video program shows the massive reflector dishes of the radio telescope array at Owens Valley in California. Radio astronomers explain how radio telescopes work and how interferometry enhances the resolution of radio telescopes.

The final segment describes how space-based telescopes enable astronomers to study the higher-energy wavelengths, such as gamma rays and X rays, that cannot be observed from Earth. The Extreme Ultraviolet Explorer satellite, with its grazing incidence optics, is shown, and its operation is explained.

As you watch the video program, consider the following questions:

1. What are the principal ways in which astronomers observe the stars?

2. What are the limitations of optical telescopes and how are they being overcome?

3. How do radio telescopes gather data and what type of observations are made with them?

4. What kinds of observations do astronomers make with space-based telescopes?

Review Activities

Matching

Match the terms listed below with the definitions that follow. Check your answers with the Answer Key and review any terms you missed.

_d_ 1. electromagnetic radiation

_e_ 2. wavelength

_c_ 3. photon ? quantum of energy

_f_ 4. atmospheric window

_m_ 5. refracting telescope

_q_ 6. focal length

_L_ 7. objective ? First Image

_O_ 8. eyepiece

_P_ 9. chromatic aberration

_n_ 10. reflecting telescope

_i_ 11. prime focus ? First Focus   objective

_K_ 12. Cassegrain focus ? Bottom

_J_ 13. Newtonian focus ? side.

_H_ 14. coudé focus ? Does't move

_G_ 15. active optics

_a_ 16. equatorial mounting

_b_ 17. alt-azimuth mounting

_X_ 18. light-gathering power

_Z_ 19. resolving power

_W_ 20. seeing

_Y_ 21. magnifying power

_S_ 22. charge-coupled device (CCD)

_V_ 23. spectrograph

_U_ 24. grating

_T_ 25. photometer

_r_ 26. radio interferometer

a. mounting in which one axis points to a celestial pole

b. mounting in which one axis points vertically upward

c. a quantum of electromagnetic energy

d. fields that transfer energy, such as light waves, radio waves, and gamma rays

e. inversely related to the energy of a photon

f. wavelength region in which a particular photon can reach the ground from space

g. computerized control of the shape of a mirror

h. a focus location that does not move with the telescope

i. the point at which the objective mirror forms an image

j. a focus location at the upper side of a telescope

k. a focus location directly beneath the objective mirror

l. the lens or mirror that first produces a focused image, as opposed to modifying, enlarging, or redirecting the image

m. a telescope that has a lens for an objective

n. a telescope that focuses light with a mirror

o. a lens for visually enlarging a focused image

p. a color distortion peculiar to simple lenses but not to mirrors

q. the distance between an objective and its focused image of a very distant object

r. combining of signals from two or more telescopes to increase resolving power

s. an electronic device for recording faint images

t. a device for measuring the intensity of light

u. an alternative to a prism as a way of dispersing light into its component colors

v. a device for separating light by wave lengths

w. a term for the quality of atmospheric conditions on a given night

x. the ability of a telescope to make starlight brighter

y. the ability of a telescope to make an image larger

z. the ability of a telescope to reveal fine detail

Completion

Fill each blank with the most appropriate term from the list for that paragraph. A term may be used once, more than once, or not at all. Check your answers with the Answer Key and review when necessary.

1. The energy of a photon is related to its wavelength. A photon with a short wavelength has _greater_ energy than a photon with a longer wavelength. For example, in comparison to an ultraviolet photon, an infrared photon has _lesser_ energy. One problem faced by astronomers is that only a small portion of electromagnetic radiation reaches Earth. Earth's uppermost atmosphere absorbs

Gamma rays, _X rays_, and some _radio_ waves. About 30 kilometers above Earth's surface, a layer of ozone absorbs _ultraviolet_ radiation. In the lower atmosphere, water vapor absorbs _infrared_ radiation.

gamma	radio
greater	ultraviolet
infrared	X rays
lesser	

2. In a refracting telescope, the objective is a _lens_. The lens closest to the object being viewed is called the _objective_ lens. The light-gathering power and resolving power of a telescope depend on the size of the objective lens; with a larger diameter lens, those powers are _greater_. In a telescope, when observing visually, the image is magnified by the _eyepiece_. One of the main problems of refracting telescopes is the inability to focus all colors of light simultaneously, which is known as _chromatic_ aberration.

chromatic	lesser
eyepiece	mirror
greater	objective
lens	

3. In a reflecting telescope, the objective is a _mirror_. The image is formed at the prime focus at the _upper_ end of the tube and then reflected by the _secondary mirror_ to a place where it can be viewed more easily. One type of reflecting telescope, a Schmidt camera, can correct distortions in an image with a thin _lens_. One of the principal advantages of reflecting telescopes is that light does not pass through a significant thickness of _glass_.

glass	mirror
lens	secondary mirror
lower	upper

4. In a radio telescope, the function performed by the mirror in a reflecting telescope is often performed by a _Q - H_ reflector. In comparison to other types of telescopes, the resolving power of a radio telescope is _Poorer_. Astronomers overcome this problem by linking radio telescopes together to form a radio _Interferometer_. In such an arrangement, the resolving power increases to that of a single telescope with a diameter _equal_ to the distance between the dishes.

antenna	equal
better	interferometer
dish	poorer

Self-Test

Select the one best answer for each question. Check your answers with the Answer Key and review when necessary.

1. Of the types of electromagnetic radiation listed below, the one with the shortest wavelength is

   a. X ray.

   b. radio wave.

   c. visible light.

   d. ultraviolet light.

2. In addition to visible light, Earth's atmosphere is transparent to

   a. X rays.

   b. radio waves.

   c. gamma rays.

   d. infrared radiation.

3. The 200-inch Hale Telescope on Mount Palomar is

   a. the largest refracting telescope.

   b. one of the largest refracting telescopes.

   c. the largest reflecting telescope.

   d. one of the largest reflecting telescopes.

4. The newest, largest ground-based optical telescopes are supported by

   a. equatorial mountings.

   (b.) alt-azimuth mountings.

   c. stationary mountings with moving coelostat mirrors.

5. For an optical telescope located on Earth under normal seeing conditions, the main value of making ever larger telescopes is improved

   a. magnifying power.

   b. resolving power.

   (c.) light-gathering power.

6. If the eyepiece of a telescope is replaced with an eyepiece with a focal length twice as long, the magnifying power is

   a. four times greater.

   b. two times greater.

   c. four times less.

   (d.) two times less.

7. A device that measures the brightness and color of an individual star is called

   a. an interferometer.

   b. a camera.

   c. a photometer.

   d. a spectrograph.

8. A device that disperses light into its component colors using thousands of microscopic parallel grooves is called a

   a. prism.

   (b.) grating.

   c. CCD.

   d. lens.

9. The advantages of a radio telescope over a large reflecting optical telescope include all of the following **except** the ability to

    a. reveal finer detail.

    b. detect cool hydrogen.

    c. see through interstellar dust.

    d. see some of the most distant objects in the universe.

10. The main reason for creating an interferometer with two separate radio dishes is to

    a. monitor two different radio wavelengths simultaneously.

    b. increase the intensity of the detected signal.

    c. compare objects in two different directions simultaneously.

    d. increase the resolution in the data.

11. A high-altitude airplane is used to make observations

    a. at gamma-ray wavelengths.

    b. at X-ray wavelengths.

    c. in the ultraviolet.

    d. in the infrared.

12. The Hubble Space Telescope is

    a. a gamma-ray telescope.

    b. an X-ray telescope.

    c. an optical telescope.

    d. a radio telescope.

Answer Key

Matching

1.	d	14.	h
2.	e	15.	g
3.	c	16.	a
4.	f	17.	b
5.	m	18.	x
6.	q	19.	z
7.	l	20.	w
8.	o	21.	y
9.	p	22.	s
10.	n	23.	v
11.	i	24.	u
12.	k	25.	t
13.	j	26.	r

Completion

1. greater, lesser, gamma, X rays, radio, ultraviolet, infrared
2. lens, objective, greater, eyepiece, chromatic
3. mirror, upper, secondary mirror, lens, glass
4. dish, poorer, interferometer, equal

Self-Test

(Page numbers refer to the textbook.)

1. a (Objective 1; page 82)
2. b (Objective 1; pages 82-83)
3. d (Objective 2; pages 86–87; video program)
4. b (Objective 2; page 90)
5. c (Objective 3; pages 90–91; video program)

6. d (Objective 3; page 93)

7. c (Objective 4; pages 95–96)

8. b (Objective 4; page 94)

9. a (Objective 5; pages 97–100; video program)

10. d (Objective 5; page 98; video program)

11. d (Objective 6; page 101)

12. c (Objective 6; pages 103–106)

Unit I

Background Notes

Scientific Method

One of the great challenges in the science of astronomy is to study objects and phenomena at very great distances from us. Unlike biology or zoology, astronomers usually cannot bring what they study into the laboratory for closer analysis. Therefore, in their observations and investigations, it is critical for astronomers to follow a consistent and logical process in developing explanations for the phenomena being studied. Adherence to this process is vital in all sciences because it allows scientists and the general public to have confidence that the knowledge gained describes the physical universe as it really is, to the best of human knowledge at a given point of time. This process not only gives us confidence in our understanding of nature, but also allows us to discover laws that govern the behavior of nature. This logical and consistent process is called the scientific method.

In the scientific method, investigation or study of a particular object or phenomenon begins with observation. Scientists record information about the subject under study. For astronomers, this information may be in the form of numerical data or photographic images made from visible, X-ray, or infrared light. Because of technology, astronomers don't have to see with their eyes everything they measure.

After completing their observations, scientists organize their findings and use them to develop an explanation or proposal to account for the phenomenon being studied. This initial explanation is called a **hypothesis**. A hypothesis must be worded in such a way that it permits other scientists to test the explanation. Scientists conduct experiments and acquire additional observations to see if a hypothesis is reasonable. A hypothesis should also be such that it can be used to predict new observations and the results of new or different experiments. Furthermore, a hypothesis cannot violate or contradict already known and accepted facts. It must be consistent with what has already been validated. Even though a hypothesis may break new ground or introduce a new way of looking at things, it cannot disagree with what is already verified. A hypothesis must expand and enhance knowledge.

If a hypothesis is tested extensively and found to be true, it may be elevated to the status of a **model** or **theory**. A model is a picture of a natural phenomenon. For example, the model of the atom as an object consisting of a nucleus with electrons in orbit about it is a picture of the smallest structure in the universe. Although, at this time, we are not certain this "picture" of the atom is correct, the model enables us to understand and predict the atom's behavior. In contrast to a model, a theory is a set of rules that describes a particular phenomenon. Also, a theory can be applied to a large number of other situations and predict behavior. A theory is not limited to one event or object. To be accepted, a theory has to be tested, evaluated, expanded, and made more general.

Once a model or theory has been subjected to extensive testing and experimentation, scientists can begin to feel that they know the "truth" about an object or phenomenon. This truth is not absolute. It is probably better to say that the scientist has great confidence in what is known, or that the information in the model or theory represents the best of our present knowledge. The scientist must also be aware that the scientific method will be used to investigate further and that this "truth" may change and our best knowledge will change. This ongoing search for "truth" is the process of science and how we learn about the universe.

The Roots of Astronomy

Astronomy has its origin in that most noble of all human traits, curiosity. Just as modern children ask their parents what the stars are and why the moon changes, so did ancient humans ask themselves these questions. The answers, often couched in mythical or religious terms, reveal a great reverence for the order of the heavens.

The study of the astronomy of ancient peoples, **archaeoastronomy,** came to public attention in 1965 when Gerald Hawkins published a book called *Stonehenge Decoded.* He reported that Stonehenge, the prehistoric ring of stones, was a sophisticated astronomical observatory.

Stonehenge, standing on Salisbury Plain in southern England, was built in stages from about 2800 B.C. to about 1075 B.C., a period extending from the late Stone Age into the Bronze Age. Though the public is most familiar with the massive stones of Stonehenge, those were added late in its history. In its first stages, Stonehenge consisted of a circular ditch slightly larger in diameter than a football field with a concentric bank just inside the ditch and a long avenue leading away toward the northeast. A massive stone, the Heelstone,

stood then, as it does now, outside the ditch in the opening of the avenue.

As early as A.D. 1740, the English scholar W. Stukely suggested that the avenue pointed toward the rising sun at the summer solstice, but few accepted the idea. More recently, astronomers have recognized significant astronomical alignments at Stonehenge. Seen from the center of the monument, the summer solstice sun rises behind the Heelstone. Other sight lines point toward the most northerly and most southerly risings of the moon.

The significance of these alignments has been debated. Hawkins claims that the Stone Age people who built Stonehenge were using it as a device to predict lunar eclipses. Others have been more conservative, but the truth may never be known. The builders of Stonehenge had no written language and left no records of their intentions. Nevertheless, the presence of solar and lunar alignments at Stonehenge and at many other Stone Age monuments dotting England and Europe shows that so-called primitive peoples were paying detailed attention to the sky. The roots of astronomy lie not in sophisticated science and logic but in human curiosity and wonder.

The early inhabitants of North America were also interested in astronomy. The Big Horn Medicine Wheel (*medicine* is used here to mean magical power), on a 3000-meter shoulder of the Big Horn Mountains of Wyoming, was used by the Plains Indians about A.D. 1500–1750. This arrangement of rocks marks a 28-spoke wheel about 27 meters in diameter. Piles of rocks, called cairns, mark the center and six locations on the circumference. The astronomer John Eddy has discovered that these cairns mark sight lines toward a number of important points on the eastern horizon and could have been used as a calendar to help schedule hunting, planting, harvesting, and celebrations.

Many other American Indian sites have astronomical alignments. More than three dozen medicine wheels are known, although most do not have the sophistication of the Big Horn Medicine Wheel. The Moose Mountain Wheel, for instance, seems to have been in use as early as A.D. 100. Alignments at some Mound Builder sites of the Midwest show that these peoples were familiar with the sky. In the Southwest, the Pueblo Indian ruins in Chaco Canyon, New Mexico, among others, have alignments that point toward the rising and setting of the sun at the summer and winter solstices.

Primitive astronomy also flourished in Central and South America. The Mayan and Aztec empires built many temples aligned with the solstice rising of the sun and with the extreme points where Venus rose and set. The Caracol temple in the Yucatán is a good example. It is a circular tower containing complicated passageways and a spiral staircase (thus the name *Caracol*— "the snail's shell"). The tubelike windows at the top of the tower point toward the equinox sunset point, and the most northerly and most southerly setting points of Venus. Unfortunately, only about one-third of the tower top survives, so the directions of any other windows are forever lost.

Archaeoastronomers are uncovering the remains of ancient astronomical observatories around the world. Some temples in the jungles of Southeast Asia, for instance, are believed to have astronomical alignments.

Other scholars are looking not at temples, but at small artifacts from thousands of years ago. Scratches on certain bone and stone implements seem to follow a pattern and may be an attempt to keep a record of the phases of the moon. Some scientists contend that humanity's first attempts at writing were stimulated by a desire to record and predict lunar phases. Archaeoastronomy is uncovering the earliest roots of astronomy and simultaneously revealing some of the first human efforts at systematic inquiry. The most important lesson of archaeoastronomy is that humans don't have to be technologically sophisticated to admire and study the universe.

One thing about archaeoastronomy is especially sad. Although we are learning how primitive people observed the sky, we may never know what they thought about their universe. Many had no written language. In other cases, the written record has been lost. Dozens of beautiful Mayan manuscripts, for instance, were burned by Spanish missionaries who believed they were the work of Satan. That is one reason why our history of astronomy really begins with the Greeks. Some of their manuscripts have survived, and we can discover what they thought about the shape and motion of the heavens.

Michael A. Seeds, *Foundations of Astronomy*, 1994 ed. [Belmont, Calif.: Wadsworth, 1994], Reprinted by permission of the publisher.

The Foucault Pendulum

In 1851 the French physicist Jean Foucault provided an unambiguous demonstration that Earth rotates on its axis. Foucault suspended a 25-kilogram mass on a 60-meter wire from the domed ceiling of the Pantheon in Paris. He started the pendulum swinging and soon noticed that the direction of the pendulum's swing, the "plane" of the pendulum's motion, was changing. The pendulum continued to swing back and forth, but the direction in which the pendulum was swinging changed. In other words, the pendulum was turning.

Foucault knew that the only force acting on the pendulum was gravity, and this force could not cause the pendulum to turn. He concluded that the plane of the pendulum was not turning but that Earth was rotating under the pendulum. What appeared to be the shifting of the direction of pendulum's plane of motion was actually the room and Earth turning beneath the pendulum.

Eratosthenes' Calculation of Earth's Circumference

Eratosthenes was one of the ancient Greeks who first applied logic and reasoning to explain and understand the natural world. The method he used for determining the circumference of Earth is a classic example of scientific reasoning.

Eratosthenes was an astronomer. Around 200 B.C., he used simple rules of geometry to develop a method of determining the circumference of Earth and hence its radius. The method Eratosthenes used was as follows:

He observed that the sun was directly overhead and shone down deep vertical wells in the ancient city of Syene, Egypt, at noon on the first day of summer. He also measured that the sun was about 7 degrees south of the zenith at the same time when viewed from his home in Alexandria, several hundred miles north of Syene. Eratosthenes knew that 7 degrees was roughly one-fiftieth of a circle. By applying geometry, he showed that the distance between the cities of Alexandria and Syene was approximately one-fiftieth of the circumference of Earth. A multiple of that distance gave him the approximate circumference of Earth. (See diagram on opposite page.)

In Eratosthenes' time, the distance between Alexandria and Syene was measured at 5,000 stadia. A stadium is an ancient Greek unit of distance, roughly equal to 1/6 kilometer. Although we are not certain of the *exact* length of the stadium, Eratosthenes' calculation (50 x 5,000 stadia = 250,000 stadia) produced a result for the circumference of Earth that is within 5 percent of today's accepted value.

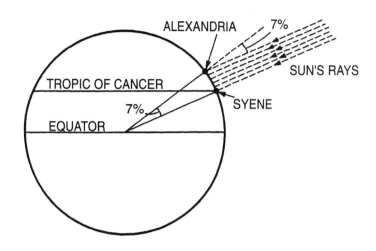

ERATOSTHENES' MEASUREMENT OF THE CIRCUMFERENCE OF EARTH
250,000 STADIA X 1/6 KILOMETER/STADIUM
= 42,000 km

The accepted value today for Earth's circumference as measured by orbiting spacecraft is 40,070 kilometers. When one considers that Eratosthenes made his calculation using only a single linear unit of measure (the stadium) and his observation of a natural phenomenon, his accomplishment is quite impressive. He had no calculators, computers, or observational satellites.

You may be curious as to why Eratosthenes calculated the circumference of Earth as opposed to the length and width of a flat surface. Greek scientists had known that Earth was round since at least the fourth century B.C. They had observed that a ship's mast slowly sank below the horizon as the ship sailed out to sea. If Earth was flat, they reasoned, the mast would disappear abruptly. Furthermore, the shadow of Earth during eclipses of the moon was also observed to be round, thus indicating that the object producing the shadow was a sphere.

The Law of Universal Gravitation

Newton's law of universal gravitation was a simple statement, based on his observation of the moon and of objects on Earth, that all objects attract all other objects. The force of that attraction is proportional to the masses of the objects and inversely proportional to the square of the distance between them. When Newton put forth his law of universal gravitation in his *Principia*, other scientists wondered about how a force could act on objects at very great distances. Newton's concept of gravity seemed almost mystical, not scientific. Newton made no effort to explain what gravity was, only to state that the law worked. The law did work in describing how the moon orbited Earth and how objects fell to the surface of Earth. However, over the next hundred years, scientists began to use the law of gravity to describe the motion of objects at great distances and began to demonstrate that the law appeared to be universal.

The first such demonstration was conducted by Edmund Halley in the late 1600s. In studying astronomical records, Halley noticed that the orbits of several bright comets that appeared in 1531, 1607, and 1682 were similar. Because of the similarity of the orbits, Halley believed that these were appearances of a single comet. He used the newly published law of universal gravitation to predict the return of this comet. His prediction came true in 1758, 16 years after his death. We now know that this comet has been observed at every return since 240 B.C.; its most recent appearance was in 1985/86. Comet Halley was at every return named in honor of the first man to calculate its orbit.

Approximately 80 years after the first predicted return of Comet Halley, John Couch Adams and Urbain Leverrier used the law of universal gravitation to explain the peculiar motions of the recently discovered sixth planet, Uranus. Uranus, discovered accidentally by William Herschel in 1781, was observed not to follow exactly the orbit predicted by Newton's law of universal gravitation. Scientists understood that gravitational forces other than the sun acted on Uranus. These additional forces are referred to as gravitational perturbations. But even after taking into account the effects of the known planets, Uranus did not follow the orbit predicted by Newton's laws. Leverrier and Adams suggested that another planet beyond the orbit of Uranus was causing the extra perturbations. They independently calculated the position of the unknown planet. Their calculations proved to be correct when Neptune was discovered in 1846. Universal gravitation applies at distances of at least 30 astronomical units.

The first observation that the force of gravity worked beyond the solar system and between stars was made by William Herschel in 1804. He noticed that the fainter companion of the star Castor moved slightly relative to its brighter component. Of course, it required the application of Newton's law of universal gravitation to show that the gravity between stars was the same as the force between planets in our solar system. That proof came in 1827 when Felix Savary was able to show that the orbits of the stars in the binary system XI Ursae Majoris were ellipses. With the determination of the 60-year period of their orbits about each other, Newton's law of gravitation became truly universal.

Special and General Relativity

Introduction

These background notes are provided to help you understand the origin, postulates, and consequences of the special and general theories of relativity. One of the difficulties in understanding these laws is that, except for a few scientists, most people have never had any personal experience with the conditions that relativity explains. No aircraft or spacecraft has ever traveled fast enough for the consequences of special relativity to become significant. We have never been in a gravitational field so strong that our motion had to be described by general relativity. Because we lack personal experience with these theories, the results or consequences these theories predict cannot be arrived at by using our common sense. In other words, special and general relativity are counterintuitive because they seem to defy common sense. But, think again. Why should nature have to follow our expectations, which are based on our experiences, in areas in which we have never personally participated?

Although the laws of relativity are counterintuitive, it is possible to understand them. One way to learn these topics is to approach them from a historical point of view. The following notes will develop the history surrounding the evolution of the special and general theories of relativity. Through this history, we can see what problems arose during the study of certain events and hypotheses. We can learn what individuals were involved in the efforts and what contributions and missteps these individuals made. By following the experiments that have been conducted, we can see the physical evidence that supports the special and general theories of relativity. Although these theories require some mental gymnastics to understand, we can be confident that substantial experimental evidence supports these laws. Remember, in order for these theories to be useful, they must explain related events and predict new phenomena to observe.

Searching for the Ether

From the time of Aristotle through the end of the nineteenth century, scientists and philosophers wondered about the nature of the material that existed between Earth and the sun. They believed light needed something through which to travel in its journey from the sun to Earth. It was difficult, if not impossible, to conceive that light could travel such great distances through a void. Thus "something" had to exist in that space. Aristotle devised the notion of the ether, and its existence became a central component of nineteenth-century science. Its existence was made even more necessary when experiments were conducted in the 1840s and 1850s that proved that light had wavelike properties. The logic went something like this: If sound waves require air through which to travel and water waves require water through which to travel, then light waves must require a material (ether) through which to travel.

The existence of the ether was so well accepted that its properties were well defined. The speed of a wave depends on the stiffness of the medium through which it travels, and since light is the fastest of all waves, the ether must be extraordinarily stiff or simply incompressible. Furthermore, Earth was not slowed in its journey around the sun because of the ether. Therefore, the ether had no viscosity. It was a perfectly transparent fluid. Based on these properties, the only thing left to do was to detect the existence of the ether. The results of these experiments to detect the ether revealed cracks in Newton's classical laws of physics.

Michelson-Morley Experiment

The most famous and accurate of all the experiments to detect the ether was conducted by Albert A. Michelson and Edward R. Morley at the Case Institute of Technology in Cleveland, Ohio, in 1887. The experiment consisted of splitting a beam of light into two individual beams that traveled at right angles (perpendicular) to each other along the arms of an instrument called an interferometer. The arms of the interferometer were of exactly equal length and had a small mirror at each end. The beams of light reflected off these mirrors and returned, traveling exactly the same distance, into an eyepiece to be studied. The beams overlapped and created an interference pattern of light and dark circles. If the beams came back at exactly the same time, the central circle would be bright. If one beam trailed behind the other by the even smallest amount, the central circle would be dark. This instrument could detect the difference in arrival times of one part in 100 million.

Since Earth was believed to move through the ether, the hypothesis was that the beam of light moving in the direction of Earth's motion around the sun would be affected by the ether and lag behind the beam traveling perpendicular to Earth's motion. (The experiment was similar to a race between airplanes, in which one plane had to fly with both a very stiff head wind and a tail wind. The other plane, flying across the wind, would always win the race.) In the Michelson-Morley experiment, the expected results were a dark circle in the center of the interference pattern. The

experiment was conducted thousands of times. The interferometer was rotated through a variety of angles and pointed in different directions. The results were always the same and opposite to what had been expected: a bright central circle in the pattern. The experiment had failed to detect the existence of the ether. Additional experiments conducted with better instruments over the next several decades all ended with the same results: The ether could not be detected. The conclusion of these findings was that the speed of light had to be unaffected by the motion of Earth. Thus, the speed of light was the same in all directions. Even though Michelson was the first American to win the Nobel Prize in physics, for his scientific instrumentation, he considered this experiment a failure and lamented it for the last 50 years of his life.

Numerous scientists attempted to explain the failure of the experiment. In 1892 G. F. Fitzgerald proposed that the interferometer arm that pointed in the direction of Earth's motion was contracted for some unknown reason. He actually developed an equation that related the amount of the contraction with the speed of the interferometer. Hendrik Lorentz, probably the most famous scientist at that time, believed very strongly that there must be a physical cause for the results of the experiment. He too accepted the notion that the interferometer was contracted. Lorentz believed that the contraction was due to some undetermined property of the electron, which had recently been discovered. Because all matter was made up of electrons, if the electrons were contracted in the direction of motion, so should be the interferometer.

Lorentz, however, went further. There was more to Michelson's experimental results than simply the contraction of the interferometer. He recognized that the speed of light had to be the same in all directions and for all observers. As a result, Lorentz developed the mathematics that showed not only how length contracted due to motion but also that time slowed for moving clocks. In other words, space and time were joined. It was no longer correct to think of absolute motion or absolute rest as Newton had done. The equations and theory developed by Lorentz were designed to explain the "failed" experiment of Michelson and Morley. The conclusion that the interferometer was shorter in one direction than in another seemed too obvious. The solution to the real problem was hinted at by the French mathematician Poincaré in 1899, when he suggested that a new and more fundamental law was needed. He restated Galileo's principle of relativity: it is impossible to detect absolute motion or absolute rest in the laboratory. It was Albert Einstein who provided the fundamental law in 1905 by asking a completely different question.

Einstein's Special Theory of Relativity

Einstein observed an even more fundamental problem with Newtonian physics: an inconsistency between Newton's laws of motion and James Maxwell's (1864) laws of electricity and magnetism. In order to reconcile these two sets of laws, Einstein stated two fundamental propositions in his special theory of relativity. The first holds that the laws of physics are the same for all observers in uniform motion; the second states that the speed of light is the same for all observers regardless of their motions. These two statements are enough to link the laws of electricity and magnetism with Newton's laws, explain the results of the Michelson-Morley experiment, and derive the Lorentz equations. The difference between the work of Lorentz and that of Einstein is that Lorentz was interested only in resolving the problem presented by the failure to detect the ether. The conclusion that the speed of light was the same for all observers was simply a consequence of the experimental result. On the other hand, Einstein was searching for an inner consistency between all the laws of physics; the fact that the speed of light was the same for all observers became a fundamental postulate.

It must be emphasized that the results or consequences of special relativity are just that, relative. Length contraction results in the shortening of moving objects only in the direction of their motion. Time dilation refers to the slowing of time (not merely clocks) for objects in motion relative to those clocks considered to be at rest. For example, length is contracted in a moving spacecraft by an observer who is considered to be at rest, and moving clocks run more slowly than clocks that are at rest relative to the observer.

One example of the consequences of special relativity is shown in the video program. The *muon* is a tiny particle that is created in Earth's upper atmosphere by bombarding cosmic rays. The particle is so unstable that it decays within one two-millionths of a second when produced in a laboratory. It would therefore be expected that few, if any, muons would reach Earth's surface. However, muons do reach Earth. Because of the muon's very high velocity, its "clock" runs more slowly than clocks on Earth. However, if you were to ride on a muon, the muon "clock" would, to you, appear to be running at normal speed. The muon disintegrates in one two-millionths of a second according to its clock. This sounds like a paradox. The solution depends on what the muon "sees." Its "clock" runs normally, but the muon "sees" Earth approaching it at very high velocity. The distance between the muon and the surface of Earth is contracted. The muon has less distance to travel and plenty of time to reach the surface of Earth. The muon illustrates the nature of special relativity.

Equivalence of Matter and Energy

Special relativity has several other consequences. The most relevant consequence of special relativity to the study of astronomy involves the equivalence of matter and energy. The famous equation $E = mc^2$ connects matter and energy and provides us with the understanding of how stars convert matter into energy.

The energy contained in matter is extremely large. The "c" in the equation refers to the speed of light. If you could convert 1 gram of matter (there are 28 grams per ounce), or roughly the mass of two paper clips, into total energy, you would have nearly 10^{14} joules of energy, or enough energy to supply your family's needs for more than 1,000 years. It is the conversion of matter into energy through the thermonuclear process of fusion that provides the sun with its energy. It is special relativity that provides us with the understanding of this process.

General Relativity

Special relativity is used to explain the behavior of objects that are traveling at speeds near the speed of light. Although relativity is a fundamental law that applies at all speeds, the effects are important only at high speeds. Furthermore, special relativity applies to constant speeds. Of course, not all speeds are constant. If the speed changes or if direction changes, there must be an acceleration. Einstein's general theory of relativity does more than simply extend special relativity to objects that accelerate, it provides a theory of the nature of gravity. The fundamental principle of general relativity is that acceleration is equivalent to gravity. In other words, the principle of equivalence states that it is impossible to distinguish between uniform acceleration and being at rest in a gravitational field.

Imagine the moment you step on your car's accelerator pedal. You are pressed back into the seat. Imagine the moment an elevator begins to rise upward. You feel somewhat heavier, you press a little harder on the floor. Imagine riding a roller coaster and reaching the bottom of the first drop. As you change direction and begin to move up the next rise, you feel yourself pressed hard into the seat of the roller coaster. All these sensations result from a change in velocity or acceleration. However, these sensations can also result from an increase in the gravitational force behind you. In other words, gravity can reproduce the same feelings you get when fuel is pumped into an engine or when your direction changes rapidly.

Acceleration and gravitation are equivalent. In fact, unless you saw the engine or felt the vibrations, it would be impossible to distinguish between gravitational effects and other forms of acceleration. This concept is the foundation of general relativity.

General Relativity and the Curvature of Space-Time

The study of special relativity has shown that the three dimensions of space (left and right, backward and forward, up and down) cannot be held separate from time, the fourth dimension. In fact, one cannot talk about an event unless it is located in space and time. We refer to this connection between space and time as *space-time*. Although space-time is fundamental to the nature of the universe, it is difficult to give a simple definition of space-time. Planets move through space-time. We conduct experiments in physics and astronomy in space-time. The galaxies move and collide in space-time. According to general relativity, the geometry of space-time is determined by the presence of matter. In other words, space-time is curved in the presence of matter.

If you were to fire a laser beam into space and aim it in such a way that it remains far from stars and other forms of matter, that beam of light would travel, for the most part, in a straight line. However, if the beam were to come near a large mass, that beam would have to follow along a curved path through space-time as produced by that large mass. Matter curves space-time, and matter and energy must follow along the paths of curved space-time. One way to look at curved space-time, although oversimplified, is to imagine a large flat sheet of rubber. Place a grapefruit in the middle of the sheet and the sheet bends in that area. Place a heavy sphere of iron and the sheet is even more curved. Roll a small object toward the depression in the sheet and that object will move along curved paths. The closer the object, the more curved the path. As long as the object remains on the rubber sheet, it must follow along those curved paths. (This analogy is two-dimensional, remember, space-time is four-dimensional.)

This concept of the curvature of space-time is very much different from Newton's view of gravity, that the gravitational force depends on the masses of the two objects involved. Light or other forms of energy have no "mass" and therefore, according to Newton's law of universal gravitation, should be unaffected by gravity. This is not the case. As first observed during a solar eclipse in 1919, beams of starlight were deflected around the sun, in accordance with the predictions made by general relativity. This experiment has been conducted numerous times with even greater degrees of accuracy.

Gravitational Red Shift/Time Dilation

The video program describes some of the other astronomical events explained by general relativity. The precession of the perihelion of Mercury's orbit and the bending of light around the sun are two of the most famous examples of experimental evidence that support and verify general relativity. Since gravity is said to curve space-time, then time must also be affected by gravity. Time runs more slowly in a stronger gravitational field. For example, if you place a clock at the top of a tall building and an identical one in the basement (the gravity at the top of the building is weaker than at the bottom because the top of the building is farther from the center of Earth), the clock in the basement will run more slowly than the clock on the top floor. In fact, an experiment with two atomic clocks, one at the top and one at the bottom of a tall building, has been conducted, and the results were consistent with those predicted by general relativity.

This slowing of time is an example of gravitational red shift, a phenomenon that has also been observed in the light emitted by massive stars as well as by the sun. Atoms within the sun that produce the light the sun emits can also be considered to be tiny clocks. Because the gravity of the sun and massive stars is large, the atoms emit light at a slightly lower frequency, or longer wavelength (slower vibrations) than the same atoms emitting light on Earth, which has a weaker gravitational field. Therefore, we would observe that the light from these atoms would be shifted slightly to the red end of the visible spectrum, hence the term gravitational red shift. Although difficult to measure, the slowing-of-time effect due to strong gravitational influences has been observed.

Gravity Waves

Another prediction of general relativity is the existence of gravity waves, or gravitational radiation. Imagine a boat moving through the water of a calm lake. The boat sets up waves that move through the water and may even reach the other side of the lake. By floating in the water, one could feel these waves and then realize that a boat has just passed by without even seeing the vessel. Massive objects do the same thing through space-time. As massive stars and other massive compact objects move through space-time, they create gravity waves that travel through space-time and produce ripples, just as a boat produces ripples on the water. It is theoretically possible to detect these waves and examine, in greater detail than visible light would allow, the nature of these objects. As of this date, however, gravity waves have not been directly detected.

Although gravity waves have not been directly detected, their existence has been implied by a number of experiments. The most famous experiment began in 1974 when Russell Hulse and Joseph Taylor discovered a binary (double) star system that consisted of two neutron stars, or pulsars. These stars, very dense and compact objects about ten miles across, emit pulses of radio energy at intervals so precise that they can be used as very accurate clocks. The period of the binary star system was measured to be 7 hours and 45 minutes. (An amazingly short time, considering that it takes Earth one year to orbit the sun.) According to

general relativity, the orbit of the binary system should be decaying or decreasing in size as the result of the emission of gravity waves. The first test of this hypothesis, reported in 1978, successfully confirmed that the system was decaying according to the predictions of general relativity. Additional experiments were conducted with even better results. In the final analysis, gravity waves have yet to be detected directly, but their existence is all but confirmed by the results of the experiments of Hulse and Taylor, who were honored for their work with the 1993 Nobel Prize in physics.

General relativity is the best existing explanation for the nature of gravity and the behavior of objects that travel through space-time. General relativity not only describes the behavior of individual objects, but also the nature of the entire universe. It is the application of general relativity that makes it possible to describe the expanding universe (in fact, general relativity predicted the expansion, although Einstein refused to believe it) and to determine the requirements needed to predict its destiny.

Atoms and Starlight

Assignments

For the most effective study of this lesson, we suggest that you complete the assignments in the sequence listed below.

Before Viewing the Video Program

- Read the Overview and Learning Objectives for this lesson. Use the Learning Objectives to guide your reading, viewing, and thinking.
- Read Chapter 6, "Atoms and Starlight," pages 111–131, in the textbook.
- Read the Viewing Notes in this lesson.

View "Atoms and Starlight" Video Program

After Viewing the Video Program

- Briefly note your answers to questions listed at the end of the Viewing Notes.
- Review all reading assignments for this lesson, especially the textbook Chapter 6 summary on page 130 and the Viewing Notes in this lesson.
- Write brief answers to all the review questions at the end of textbook Chapter 6 to be certain you understand the test material.
- Complete the Review Activities in this guide to reinforce your understanding of important terms and concepts. Check your answers with the Answer Key at the end of this Lesson and review when necessary.
- Take the Self-Test in this guide to measure your achievement of the Objectives. Check your answers with the Answer Key and review when necessary.
- Complete any other activities and projects assigned by your instructor.

Overview

An earlier lesson, "The Tools of Astronomy," introduced the principal devices used in astronomical study and research. After the telescope, the spectrograph is probably the single most important tool astronomers have. For it is a spectrograph that enables astronomers to dissect the only "material" the distant stars provide: light. A spectrograph spreads out the light so astronomers can see what they have.

Although the concepts in this lesson may seem abstract, they are important. Since astronomers have only light to work with, an understanding of the nature of light and of how it interacts with matter is essential to drawing correct conclusions about the source of that light, whether from a planet, star, nebula, or galaxy.

The spectrograph works by dispersing light into its component colors. The two basic types of spectrographs use either a prism, a wedge-shaped piece of glass, or a grating, a piece of glass with thousands of fine parallel lines scribed into its surface. Once a spectrograph has dispersed light into its component colors, astronomers can extract an impressive amount of information from a spectrum.

Detailed analysis of a stellar spectrum enables astronomers to identify the material composing a star even though they have none of that star's material in their possession. The stellar spectrum also provides information about the temperature of a star's surface layers. In fact, the close similarity between the spectrum of the sun and those of many other stars was one of the first indications that our sun *is* a star.

Spectra also provide information about a phenomenon known as the Doppler effect, which is a change in the wavelength of radiation that results from the relative motion of the source of light and the observer of that light. The Doppler effect can reveal if a star is moving toward or away from Earth; its speed; its rate of rotation and whether it is expanding or contracting, exploding or being bombarded by infalling material, calm or turbulent. The Doppler effect can even be used to determine if there are actually two stars orbiting each other, even though telescopes see only a single point of light.

Learning Objectives

When you have completed all assignments for this lesson, you should be able to:

1. Describe a simplified model of a typical atom, with the arrangement of its subatomic particles and energy levels, and explain how such an atom interacts with light. TEXTBOOK PAGES 113–116; VIDEO PROGRAM.

2. Describe the appearance, origin, and characteristics of absorption, emission, and continuous spectra. TEXTBOOK PAGES 116–120; VIDEO PROGRAM.

3. Describe the origin of the first three series of lines in the spectrum of hydrogen and of 21 centimeter radiation. TEXTBOOK PAGES 120–123 and BOX 6.3, PAGES 124–125.

4. Explain spectral classification and how the appearance of a spectrum indicates stellar temperature. TEXTBOOK PAGES 123–127; VIDEO PROGRAM.

5. Explain what spectra indicate about stellar chemical composition and about stellar motions as revealed by the Doppler Effect. TEXTBOOK PAGES 128–129; VIDEO PROGRAM.

Viewing Notes

The opening segment of "Atoms and Starlight" traces how the scientific study of light evolved from Newton's experiments with optics through the work of Joseph Fraunhofer in the early nineteenth century, who discovered dark lines in the sun's spectrum, to the work of Robert Bunsen and Gustav Kirchhoff, two German chemists who discovered the significance of those lines. The segment describes the experiments Bunsen and Kirchhoff conducted that enabled them to identify the three basic types of spectra: continuous, emission, and absorption. Computer graphics illustrate how each spectrum is produced and relate them to different celestial phenomena.

The video program then examines how physicists, after learning *how* light behaves came to understand *why* light behaves in its special ways. As this segment explains, the lines in a spectrum result from the nature of atoms and how atoms interact with light. Astronomers, aided by computer graphics, supply detailed descriptions of atoms and their electrons and how transitions of electrons between energy levels produce photons of different wavelengths.

It is the photons that astrophysicists actually study when they unlock the messages of the stars, and the next segment, "Spectroscopy," explores the tools and techniques of this science. The video program shows modern spectrographs, the instruments that enable astronomers to determine a star's composition. Two spectrographs featured in the program are the coudé spectrograph at Kitt Peak National Observatory and the fiber-optic spectrograph at Palomar Mountain Observatory. The coudé spectrograph is so large a person can walk around inside it, and the fiber-optic spectrograph can obtain the spectra of dozens of stars simultaneously by using fibers to position the light from each star along the spectrograph's slit. At both locations, observatory staff explain the operation of the instruments.

In addition to enabling astronomers to determine a star's chemical composition, the spectrum supplies information that can be used to determine a star's temperature. In the next segment, astronomers explain how the strength of spectral lines is related to stellar temperatures. Computer animation illustrates the modern spectral classification system—O, B, A, F, G, K, M—and its relationship to stellar temperatures. The video program concludes by explaining how, through the Doppler effect, a star's spectrum reveals clues about its motion with respect to Earth and its rotation rate.

As you watch the video program, consider the following questions:

1. What are the basic parts of an atom? How does an electron absorb or emit a photon of light?

2. What are the three different kinds of spectra, and what kinds of objects (either astronomical phenomena or manufactured objects) produce them?

3. How and why can a spectrum be analyzed to determine a star's temperature?

4. How and why can a spectrum be analyzed to determine a star's chemical composition?

5. How and why can a spectrum be analyzed to determine a star's motion?

Review Activities

Matching

Match the terms listed below with the definitions that follow. Check your answers with the Answer Key and review any terms you missed.

_j_ 1. nucleus

_v_ 2. proton

_x_ 3. neutron

_w_ 4. electron

_u_ 5. isotope

_e_ 6. ionization

_o_ 7. ion

_b_ 8. molecule

_n_ 9. Coulomb force

_f_ 10. binding energy

_L_ 11. permitted orbit

_K_ 12. energy level

_r_ 13. excited atom

_q_ 14. ground state

_C_ 15. black-body radiation

_d_ 16. continuous spectrum

_i_ 17. absorption line

_y_ 18. absorption spectrum

_z_ 19. emission line

_G_ 20. emission spectrum

_a_ 21. Kirchhoff's Laws

_m_ 22. transition

_s_ 23. Lyman, Balmer, and Paschen series

_p_ 24. spectral class

_t_ 25. Doppler effect

_H_ 26. radial velocity

a. rules governing the absorption and emission of light by matter

b. two or more atoms bonded together

c. energy emitted by a hypothetical perfect radiator

d. the kind of spectrum produced by a perfect radiator

e. the process by which atoms lose or gain electrons

f. the energy needed to pull a particular electron away from its atom

g. the kind of spectrum produced by a low-density gas

h. rate at which an object approaches or recedes from the observer

i. a dark-line in a spectrum

j. the central core of an atom

k. the state of an electron with a particular binding energy

l. any energy level that an electron may occupy within an atom

m. movement of an electron between two energy levels

n. electrostatic repulsion or attraction between charged bodies

o. an atom that has lost or gained one or more electrons

p. an indication of a star's surface temperature based on its spectrum

q. the lowest permitted energy level in an atom

r. an atom in which at least one electron is not in the ground state

s. some of the lines in the hydrogen spectrum

t. the shift in the wavelength of spectral lines due to radial velocity

u. atoms with the same number of protons but a different number of neutrons

v. a positively charged particle in the nucleus of an atom

w. low-mass atomic particle that carries a negative charge

x. atomic particle with no electrical charge

y. the kind of spectrum produced by the sun

z. a bright line in a spectrum

Completion

Fill each blank with the most appropriate term from the list for that paragraph. A term may be used once, more than once, or not at all. Check your answers with the Answer Key and review when necessary.

1. The spectrum produced by glowing high-density gas is a _Continuous_ spectrum, and its color is determined primarily by its _temp_. The spectrum produced by glowing low-density gas is a _Bright lin_ spectrum, and its color is determined primarily by its _Chem Comp_. A normal star produces a _Dark line_ spectrum, its color is determined primarily by its _temp_, and the pattern of lines is determined primarily by its _Chem_. The relative strengths of different lines in the pattern are determined primarily by _temp._.

    bright-line _Emission Spec_   dark-line _Absorption Spec_
    chemical composition   temperature
    continuous

2. In the center of a model of a typical atom is a _Nucleus_, which is surrounded by a cloud of _Electrons_. The two different particles that make up the center of the atom are the positively charged _Protons_ and _Neutron_, which have no charge. The chemical element represented by an atom is determined by the number of _Protons_. The isotope of the element it represents is determined by the number of _Neutrons_. An ion is an atom in which the number of _Electron_ is different from the number of protons.

    electrons   neutrons
    nucleus   protons

3. In a hydrogen atom, the higher the energy level, the _lesser_ is the difference in energy between that energy level and an adjacent level. Thus, transitions to or from a higher level involve photons of _lesser_ energies and _longer_ wavelengths than transitions to or from a lower level. In a very cool star, the number of hydrogen atoms with electrons in the second energy level would be _lesser_ than those with electrons in the first energy level. In a very hot star, the number of hydrogen atoms with electrons above the second level would be _Greater_ than those with electrons in the second level. Thus, the cooler stars produce the strongest hydrogen lines in that part of the spectrum with _Shorter_ wavelengths.

    lesser   shorter
    longer   greater

Self-Test

Select the one best answer for each question. Check your answers with the Answer Key and review when necessary.

1. When an electron in an atom absorbs a photon with an energy less than the electron's binding energy, it may
   a. move to a lower energy level.
   b. remain in the same energy level.
   c. move to a higher energy level.
   d. escape from the atom altogether.

2. If left undisturbed by collisions or photons, an electron in an excited atom will
   a. eventually move to a lower energy level.
   b. remain in the same energy level forever.
   c. eventually move to a higher energy level.
   d. eventually escape from the atom altogether.

3. The hotter a star is, the
   a. brighter and redder it will be.
   b. brighter and bluer it will be.
   c. fainter and redder it will be.
   d. fainter and bluer it will be.

4. The sun produces
   a. a continuous spectrum.
   b. an emission spectrum.
   c. an absorption spectrum.
   d. no spectrum.

5. Transitions to or from the ground state of a hydrogen atom produce spectral lines in the

   a. gamma ray part of the spectrum.

   b. ultraviolet part of the spectrum.

   c. infrared part of the spectrum.

   d. radio part of the spectrum.

6. The 21-centimeter line of hydrogen is in the

   a. gamma ray part of the spectrum.

   b. ultraviolet part of the spectrum.

   c. infrared part of the spectrum.

   d. radio part of the spectrum.

7. A star's spectrum has very weak hydrogen lines but shows strong bands from titanium oxide. This star probably

   a. has very little hydrogen.

   b. is extremely hot.

   c. is moderate in temperature.

   d. is extremely cool.

8. Which one of the following spectral classes has stars with the coolest temperature?

   a. A

   b. B

   c. M

   d. O

9. The chemical element found in the greatest abundance in all normal stars is

   a. carbon.

   b. helium.

   c. hydrogen.

   d. iron.

10. Because of the Doppler effect, the spectral lines of a luminous object approaching Earth will appear

    a. broadened.

    b. split into several adjacent lines.

    c. shifted slightly toward the blue.

    d. shifted slightly toward the red.

Answer Key

Matching

1.	j	14.	q
2.	v	15.	c
3.	x	16.	d
4.	w	17.	i
5.	u	18.	y
6.	e	19.	z
7.	o	20.	g
8.	b	21.	a
9.	n	22.	m
10.	f	23.	s
11.	l	24.	p
12.	k	25.	t
13.	r	26.	h

Completion

1. continuous, temperature, bright-line, chemical composition, dark-line, temperature, chemical composition, temperature
2. nucleus, electrons, protons, neutrons, protons, neutrons, electrons
3. lesser, lesser, longer, lesser, greater, shorter

Self-Test

(Page numbers refer to the textbook.)

1. c (Objective 1; pages 115, 116; video program)

2. a (Objective 1; page 116)

3. b (Objective 2; pages 117–118)

4. c (Objective 2; pages 118–120; video program)

5. b (Objective 3; page 121)

6. d (Objective 3; pages 124–125)

7. d (Objective 4; pages 126–127)

8. c (Objective 4; pages 126–127)

9. c (Objective 5; page 130; video program)

10. c (Objective 5; page 128; video program)

The Sun

Assignments

For the most effective study of this lesson, we suggest that you complete the assignments in the sequence listed below.

Before Viewing the Video Program

- Read the Overview and Learning Objectives for this lesson. Use the Learning Objectives to guide your reading, viewing, and thinking.

- Read Chapter 7, "The Sun," pages 133–149, in the textbook.

- Read Background Notes 8, "Solar Magnetic Cycle," following Lesson 13 in this guide.

- Read the Viewing Notes in this lesson.

View "The Sun" Video Program

After Viewing the Video Program

- Briefly note your answers to the questions listed at the end of the Viewing Notes.

- Review all reading assignments for this lesson, especially the textbook Chapter 7 summary on page 147, the Background Notes 8, and the Viewing Notes in this lesson.

- Write brief answers to all the review questions at the end of textbook Chapter 7 to be certain you understand the text material.

- Complete the Review Activities in this guide to reinforce your understanding of important terms and concepts. Check your answers with the Answer Key at the end of this lesson and review when necessary.

- Take the Self-Test in this guide to measure your achievement of the Objectives. Check your answers with the Answer Key and review when necessary.

- Complete any other activities and projects assigned by your instructor.

Overview

So far in this course, we have seen the long historical process by which astronomy evolved from its ancient mystical origins and became a modern science. We have viewed the ever more sophisticated telescopes, spacecraft, and auxiliary instruments that astronomers use to study the universe. And we have learned some of the basic laws of physics, relating to light and gravity for example, that govern everything in nature but that are particularly important in understanding phenomena in outer space. Yet we have not really begun to explore the astronomical phenomena themselves.

In this lesson, and for the remainder of the course, we will do just that, beginning with the sun. The sun is a fitting place to start a journey that will carry us increasingly farther into space and, in some cases, farther back in time. For the sun is a star, very similar in most respects to those myriad points of light in the nighttime sky, but closer to us than any other. It is the only star whose surface we can study in detail. Probing its secrets has certainly helped us to understand stars in general, but it works the other way too. For example, the presumption that the sun will eventually swell into a huge, intensely bright red giant star and destroy us (as is described in Lesson 12) is based on our watching other older stars undergo that very change, even though our sun has not yet done that.

Although our dependence upon the sun's light and heat is obvious, we are slowly learning that the sun is curiously undependable. Its 11-year magnetic cycle, revealed most clearly by the changing number of sunspots, has long been known to wreak havoc with delicate technologies, such as navigational and communications equipment and orbiting satellites. That cycle is poorly understood, so it is perplexing when the cycle shuts down for extended periods, as it has several times in the past few thousand years, apparently resulting in prolonged cold spells on Earth. For all of its nearness, the sun remains mysterious in ways that affect our lives.

This lesson concentrates on the sun's surface features and activity and their effect on Earth, the very kinds of phenomena that cannot be studied easily in distant stars. The next five lessons are devoted to every other aspect of stellar birth, life, and death, and include more detailed descriptions of the sun's hidden interior and nuclear reactions.

Learning Objectives

When you have completed all assignments for this lesson, you should be able to:

1. Identify and describe the three major layers of the solar atmosphere, how each is observed, and the principal phenomena located in each layer. TEXTBOOK PAGES 133–138; VIDEO PROGRAM.

2. Describe the behavior and suspected cause of the solar magnetic cycle and its effects on sunspots, prominences, flares, and coronal activity. TEXTBOOK PAGES 138–147; VIDEO PROGRAM.

3. Describe known and suspected effects of the solar magnetic cycle on Earth and its immediate environment in space. TEXTBOOK PAGES 144–145; BACKGROUND NOTES 8; VIDEO PROGRAM.

4. Describe variations from the usual solar magnetic cycle and their suspected effects on Earth. TEXTBOOK PAGES 138–140; BACKGROUND NOTES 8; VIDEO PROGRAM.

Viewing Notes

After the introduction reminds us how much the sun has in common with other stars, the first segment emphasizes how differently we study the sun than other stars. It concerns solar telescopes and related tools. Although the ultimate goal of solar and stellar observations may be the same—to study spectral features, for instance—the sun's blinding brightness and atmospheric turbulence-causing heat make equipment used to observe the sun quite unique.

The video program shows portions of the 60-foot solar tower on Mount Wilson. Observations made on Mount Wilson were used, along with the Zeeman effect, to discover the magnetic nature of sunspots early in this century. Solar astronomers and astrophysicists explain helioseismology, the study of vibrations in the visible surface of the sun that reveal the underlying hidden structure, especially the sun's rate of rotation at different depths. The advantages of studying the sun from space are also explained. From space, we avoid the turbulence of Earth's atmosphere and the atmospheric absorption of ultraviolet, X-ray, and other wavelengths, which enables us to study solar features like the corona without having to wait for a total solar eclipse and allows us to sample the solar wind. The video program mentions the pioneering work of Skylab and the Solar Maximum Mission and previews the ongoing Ulysses mission, which was sent to study the sun from above its poles, a view not possible from Earth.

In the next segment, computer graphics provide a complete inside-out view of the sun's structure. The first three layers compose the sun's interior and, although they are not described in detail by the textbook until Chapter 9, they are included briefly in the video program because the outermost layer, the convective zone, is largely responsible for what we see at the surface: granules and supergranules, sunspots and the solar wind, differential rotation, and the magnetic cycle. The video program then looks at the layers of the solar atmosphere: photosphere, chromosphere, and corona.

The final segment of the video program concerns the solar magnetic cycle and its effect on Earth and interplanetary space. It looks at sunspots, prominences, and flares in detail and uses the Babcock model to suggest that the ability of the gaseous sun to rotate at different speeds at different latitudes is somehow responsible for the 11-year magnetic cycle. The Babcock model, however, cannot account for the occasional cessation of that cycle and corresponding effects on Earth's climate. Several astronomers comment on the impact of this shutdown. The "Little Ice Age," mentioned by Libbrecht, a solar astrophysicist, is thought to have resulted from a period of reduced sunspot activity known as the Maunder minimum. (Background Notes 8 contain additional information on the Maunder minimum and how variations in the magnetic cycle affect Earth.)

As you watch the video program, consider the following questions:

1. How does our study and knowledge of the sun differ from that of other stars?

2. What does the study of other stars tell us about our sun?

3. How is our sun studied? What are the advantages of studying the sun from above Earth's atmosphere?

4. What are the three layers of the solar atmosphere?

5. What features are found in the solar atmosphere?

6. What are some of the effects of the solar magnetic cycle?

Review Activities

Matching

Match the terms listed below with the definitions that follow. Check your answers with the Answer Key and review any terms you missed.

_M_ 1. auroras

_L_ 2. Babcock model

_b_ 3. chromosphere

_c_ 4. corona

_n_ 5. coronal holes

_f_ 6. differential rotation

_a_ 7. filtergram

_k_ 8. flare

_p_ 9. granulation

_q_ 10. helioseismology

_g_ 11. Maunder butterfly diagram

_h_ 12. Maunder minimum

_a_ 13. photosphere

_k_ 14. prominence

_d_ 15. solar wind

_j_ 16. spicule

_e_ 17. sunspot

_s_ 18. supergranule

_i_ 19. Zeeman effect

a. the visible surface of the sun

b. a nearly invisible layer of gas exhibiting a brief flash of pink light during total solar eclipses

c. the solar atmosphere above the chromosphere

d. high-velocity atoms and ions streaming away from the sun

e. a cool, dark area of the sun's surface

f. the splitting of a single spectral line into multiple lines caused by a magnetic field

g. the pattern made by the distribution of sunspots over time

h. a period of reduced solar activity

i. varying rates of movement of specific parts of the same body

j. an explanation that the magnetic cycle results from progressive tangling of the solar magnetic field

k. a solar eruption that can last for months

l. a violent eruption on the solar surface

m. glowing displays in Earth's upper atmosphere caused by gusts in the solar wind

n. cooler, lower-density region of the corona

o. photograph taken in the light of a specific region of the spectrum

p. mottled appearance of bright cells in the photosphere

q. study of the sun's modes of vibration

r. flamelike structures that can last from 5 to 15 minutes

s. a large convective feature on the sun's surface; spicules appear around the edges

Completion

Fill each blank with the most appropriate term from the list for that paragraph. A term may be used once, more than once, or not at all. Check your answers with the Answer Key and review when necessary.

1. The three major layers of the solar atmosphere are the _photosphere_ _chromosphere_, and _corona_. The visible surface of the sun is the _photosphere_, which is a _thin_ layer of _low_-density gas. The two layers that can only be seen during a _solar_ eclipse are the _chromosphere_ and _corona_.

chromosphere	photosphere
corona	solar
high	thick
low	thin
lunar	

2. The sunspot cycle is related to the sun's _rotation_. The sun, however, does not rotate uniformly. The equatorial region rotates _faster_ than regions at _higher_ latitudes. Moreover, the surface of the sun rotates _faster_ than the _interior_. These variations are known as _differential_ rotation and are clearly linked to the _magnetic_ cycle.

differential	lower
exterior	magnetic
faster	rotation
higher	slower
interior	

3. Solar flares can have many effects on Earth. X-ray and _ultraviolet_ radiation reach Earth in _8 second_ and _8 min_ _increases_ the ionization in Earth's _upper_ atmosphere, which alters the _reflection_ of shortwave radio signals. Flares also eject high-energy particles, most of which reach Earth as gusts in the _solar wind_ hours or days after the flare erupts. These gusts generate currents and interact with Earth's _magnetic_ field. Interaction of these particles with atoms in the _upper_ atmosphere can produce _auroras_ and causes _compasses_ to behave erratically. Solar flares can also cause surges in high-voltage _power lines_ and create _radiation_ hazards for passengers in supersonic transports and spacecraft.

absorption	lower
auroras	magnetic
compasses	polar
decreases	power lines
eight days	radiation
eight minutes	reflection
electrical	solar wind
increases	ultraviolet
infrared	upper

Self-Test

Select the one best answer for each question. Check your answers with the Answer Key and review when necessary.

1. The visible surface of the sun is called the
    a. photosphere.
    b. chromosphere.
    c. corona.
    d. heliosphere.

2. The highest temperature in the solar atmosphere is found in
    a. the photosphere.
    b. a sunspot.
    c. the chromosphere.
    d. the corona.

3. An absorption spectrum is produced by the
    a. photosphere.
    b. chromosphere.
    c. corona.
    d. solar wind.

4. The sun rotates
    a. fastest at the poles.
    b. fastest at mid-latitudes.
    c. fastest at the equator.
    d. equally fast at all latitudes.

5. The Zeeman effect is used to measure
    a. the rotation rate of the sun.
    b. vibrations coming from sun's interior.
    c. the speed of solar wind.
    d. the strength of solar magnetic fields.

6. The corona is largest and most circular
   a. at sunspot minimum.
   b. at sunspot maximum.
   c. at random times unrelated to sunspot cycle.
   d. during the Maunder minimum.

7. Displays of auroras are most frequent
   a. at sunspot minimum.
   b. at sunspot maximum.
   c. at random times unrelated to sunspot cycle.
   d. during the Maunder minimum.

8. Strong solar magnetic activity can have all of the following effects **except**
   a. power outages.
   b. more gradual decay of spacecraft orbits.
   c. radiation hazard to astronauts.
   d. interference with radio communication.
   e. interference with compasses.

9. The Maunder minimum lasted about
   a. 4 years.
   b. 11 years.
   c. 22 years.
   d. 70 years.
   e. 400 years.

10. The Maunder minimum is thought to have
   a. caused unusually warm weather on Earth.
   b. caused unusually cool weather on Earth.
   c. increased auroral displays.
   d. had no discernible effect on Earth.

Answer Key

Matching

1.	m	8.	l	15.	d
2.	j	9.	p	16.	r
3.	b	10.	q	17.	e
4.	c	11.	g	18.	s
5.	n	12.	h	19.	f
6.	i	13.	a		
7.	o	14.	k		

Completion

1. photosphere, chromosphere, corona (any order); photosphere; thin; low; solar; chromosphere, corona (either order)

2. rotation, faster, higher, faster, interior, differential, magnetic

3. ultraviolet, eight minutes, increase, upper, reflection, solar wind, magnetic, upper, auroras, compasses, power lines, radiation

Self-Test

(Page numbers refer to the textbook.)

1. a   (Objective 1; page 134; video program)

2. d   (Objective 1; page 138; video program)

3. a   (Objective 1; pages 134–135)

4. c   (Objective 2; page 141; video program)

5. d   (Objective 2; page 139; video program)

6. b   (Objective 2; page 146; video program)

7. b   (Objective 3; pages 143–145; Background Notes 8)

8. b   (Objective 3; page 145; Background Notes 8; video program)

9. d   (Objective 4; page 140; Background Notes 8)

10. b   (Objective 4; page 140; Background Notes 8)

9

Stellar Properties

Assignments

For the most effective study of this lesson, we suggest that you complete the assignments in the sequence listed below.

Before Viewing the Video Program

- Read the Overview and Learning Objectives for this lesson. Use the Learning Objectives to guide your reading, viewing, and thinking.

- Read Chapter 8, "The Properties of Stars," pages 150–176, in the textbook.

- Read the Viewing Notes in this lesson.

View "Stellar Properties" Video Program

After Viewing the Video Program

- Briefly note your answers to questions listed at the end of the Viewing Notes.

- Review all reading assignments for this lesson, especially the textbook Chapter 8 summary on page 175 and the Viewing Notes in this lesson.

- Write brief answers to all the review questions at the end of textbook Chapter 8 to be certain you understand the text material.

- Complete the Review Activities in this guide to reinforce your understanding of important terms and concepts. Check your answers with the Answer Key at the end of this lesson and review when necessary.

- Take the Self-Test in this guide to measure your achievement of the Objectives. Check your answers with the Answer Key and review when necessary.

- Complete any other activities and projects assigned by your instructor.

Overview

The previous lesson concerned the sun—a typical star, yet the only star whose surface we can see and study in detail. This lesson is the first of five lessons that deal with all the other stars in the sky. In this lesson, we face the challenge of how to study stars that, even when viewed through the largest contemporary telescopes, remain unresolved points of light. We cannot see their surfaces; they do not appear as even tiny circular disks of measurable diameter. Because of their tremendous distance from Earth, they are too small to be magnified significantly.

What measurements can be made of a featureless point of light? In one sense, not many. We can measure the direction of that point; that is, we can pinpoint the location of the star in the sky and note any changes in that location over time. We can also measure a star's light. We can determine its total light, and we can use a spectrograph to dissect a star's light into its component colors. Although these measurements may appear to be inadequate, we can use this information to deduce many properties of stars, such as distance, luminosity, surface temperature, mass, and diameter. This lesson illustrates how astronomers have slowly pieced together the answers to the larger questions: how do stars form; how do they work; and how do they die? The answers to these questions will be explored in the next four lessons.

Learning Objectives

When you have completed all assignments for this lesson, you should be able to:

1. Define stellar parallax and parsec and describe the use of parallax to determine stellar distances. TEXTBOOK PAGES 152–155; VIDEO PROGRAM.

2. Explain how distance, luminosity, and apparent magnitude are related. TEXTBOOK PAGES 150–152, 155–157; VIDEO PROGRAM.

3. Describe the purpose and design of the Hertzsprung-Russell (H-R) diagram, including the locations of supergiant, giant, main-sequence, and white dwarf stars, and explain how the H-R diagram is used in luminosity classification. TEXTBOOK PAGES 158–163; VIDEO PROGRAM.

4. Describe the behavior and observation of visual, spectroscopic, and eclipsing binaries and explain how they are used to determine stellar masses and diameters. TEXTBOOK PAGES 163–169; VIDEO PROGRAM.

5. Describe the mass-luminosity relation and the stellar density function. TEXTBOOK PAGES 169–174; VIDEO PROGRAM.

Viewing Notes

The story of how astronomers have learned about the properties of stars—and what they have learned—forms the basis for this video program. "Stellar Properties" begins by reviewing the challenges presented in studying the stars—most of which are seen as tiny points of light at unimaginable distances. Over time, however, astronomers have been able to overcome the challenges and learn about the properties of stars through increasingly sophisticated measurements and analyses.

One of the first measurements astronomers were able to make of the stars was their distance from Earth. The basic technique used is parallax, and the video program explains how parallax is used to measure distance. Through examples, astronomers describe the size of the second of arc, the basic unit of distance used in measuring parallax. The video program also traces the history of measuring parallax, from the unsuccessful attempts of ancient Greek astronomers to the precise data now being collected by the French satellite HIPPARCOS (HIgh Precision PARallax COllecting Satellite) and the Hubble Space Telescope.

Although distance itself is not an inherent property of a star, knowledge of a star's distance is critical because, combined with measurement of a star's apparent brightness, it can be used to calculate a star's luminosity or its actual energy output. The video program illustrates the concepts of apparent brightness and luminosity and describes what astronomers have learned about the range of luminosity among stars.

Other important knowledge of stellar properties comes from spectral analysis, which provides information about a star's surface temperature. In the video program, computer animation illustrates the clear relationship between color and temperature and how that relationship is plotted on the Hertzsprung-Russell diagram.

While astronomers were learning much about temperature and luminosity, the observations themselves did not explain the reasons for the tremendous variations in temperature and luminosity among main-sequence stars. The final segment of the video program describes how the answer was found in measurements of the diameters and masses of binary stars. This discovery of the mass-luminosity relationship is fundamental to understanding how stars work.

As you watch the video program, consider the following questions:

1. How does the parallax of an object depend on its distance from the observer as well as the distance between the two points from which the observation is made (the "baseline")?

2. How are a star's apparent brightness, luminosity, and distance related?

3. What are two ways to estimate the surface temperature of a star?

4. What two stellar properties are compared in a Hertzsprung-Russell diagram, and what is the relationship between these properties for main-sequence stars?

5. What third property can cause two stars of the same temperature to have different luminosities?

7. How are mass and luminosity related for main-sequence stars?

Review Activities

Matching

Match the terms listed below with the definitions that follow. Check your answers with the Answer Key and review any terms you missed.

_____ 1. stellar parallax

_____ 2. parsec

_____ 3. luminosity

_____ 4. H-R diagram

_____ 5. main sequence

_____ 6. giant star

_____ 7. supergiant star

_____ 8. white dwarf star

_____ 9. luminosity class

_____ 10. binary star

_____ 11. center of mass

_____ 12. visual binary

_____ 13. spectroscopic binary

_____ 14. eclipsing binary

_____ 15. light curve

_____ 16. mass-luminosity relation

_____ 17. stellar density function

_____ 18. absolute visual magnitude

_____ 19. distance modulus

_____ 20. spectroscopic parallax

a. two stars orbiting each other

b. a graph comparing the luminosities and surface temperatures of stars

c. the more massive a main-sequence star is, the more luminous it is

d. the angular difference between a star's positions seen from two different locations

e. a binary star system in which the two stars are not resolvable but exhibit variations in brightness as seen from Earth f.　the region on the H-R diagram occupied by about 90 percent of all stars and for which luminosity and surface temperature are most simply related

g.　the point about which two stars move as they orbit each other, proportionally closer to the more massive star

h.　the distance at which a star would have a parallax of 1 second of arc

i.　the total amount of energy a star radiates each second

j.　a binary star system in which both stars are visible separately

k.　a description of the abundance of stars of different types in space

l.　a star about the size of Earth that is very faint though very hot

m.　a category of stars determined by the widths of lines in their spectra

n.　a graph indicating any change of brightness over time

o.　a highly luminous star with a diameter about 10 to 100 times larger than that of the sun

p.　an exceptionally luminous star with a diameter up to 1,000 times larger than that of the sun

q.　a binary star system in which stars are too close together to be visible separately; presence of two stars is revealed by Doppler shifts

r.　a method of estimating the distance of a star by comparing its apparent magnitude with its absolute magnitude as estimated from its spectrum

s.　the intrinsic visual brightness of a star

t.　the difference between the apparent and absolute magnitude of a star

Completion

Fill each blank with the most appropriate term from the list for that paragraph. A term may be used once, more than once, or not at all. Check your answers with the Answer Key and review when necessary.

1.　The farther a star is from Earth, the _smaller_ its parallax. If we could observe the stars from Mars, which is farther from the sun than Earth, all of the stars would have _larger_ parallaxes. How bright a star appears depends on its distance and _luminosity_.

age   luminosity

larger   smaller

2. The Hertzsprung-Russell diagram compares the intrinsic brightness
   of a star to its ___S-temp___. For main-sequence stars, the
   ___hotter___ they are in surface temperature, the brighter they are
   as well. If two stars have the same diameter but one is intrinsically
   fainter, it must be ___cooler___ in surface temperature also.

   cooler   hotter

   diameter   surface temperature

3.   The motions of binary stars can be analyzed to tell us the
     ___mass___ of each star. If the stars are eclipsing, we can also often tell
     the ___diameter___ of each star. For main-sequence stars, the
     ___larger___ the mass, the greater the luminosity. For main-sequence stars,
     the larger the mass, the ___smaller___ the number of such stars we find of
     that type in space.

     diameter   mass

     larger   smaller

     luminosity

Self-Test

Select the one best answer for each question. Check your answers with the
Answer Key and review when necessary.

1. A parsec is
   a. the distance that light travels in 1 second.
   b. the number of seconds in a year.
   c. the average distance between Earth and the sun.
   d. the distance of a star with a parallax of 1 second of arc.

2. A star has a parallax of 0.1 second of arc. How far is it in parsecs?
   a. 0.1
   b. 1
   c. 10
   d. 100

3. If you moved five times farther from a distant star, the star would appear to be

    a. 5 times brighter.

    b. 5 times fainter.

    c. 10 times fainter.

    d. 25 times fainter.

4. Two stars are the same distance from Earth but star "A" has an apparent magnitude of 1 and star "B" has an apparent magnitude of 6. Which of the following statements is correct?

    a. Star "A" is 5 times more luminous than Star "B."

    b. Star "A" is 100 times more luminous than Star "B."

    c. Star "B" is 5 times more luminous than Star "A."

    d. Star "B" is 100 times more luminous than Star "A."

5. Of the following stars, the one with the largest diameter is

    a. G2 Ib.

    b. G2 II.

    c. G2 III.

    d. G2 IV.

6. Of the following main-sequence stars, the most luminous is

    a. B2.

    b. A4.

    c. F1.

    d. G8.

7. Stellar diameters can be determined by observation of

    a. astrometric binaries.

    b. eclipsing binaries.

    c. visual binaries.

    d. spectroscopic binaries.

8. For a binary star system with any given orbital size, the shorter the orbital period,

    a. the more massive the stars are.

    b. the less massive the stars are.

    c. neither; orbital period is unrelated to mass.

9. The rarest spectral class of stars is

    a. K stars.

    b. B stars.

    c. M stars.

    d. O stars.

10. The most massive main-sequence stars are

    a. A stars.

    b. G stars.

    c. O stars.

    d. M stars.

Answer Key

Matching

1.	d	11.	g
2.	h	12.	j
3.	i	13.	q
4.	b	14.	e
5.	f	15.	n
6.	o	16.	c
7.	p	17.	k
8.	l	18.	s
9.	m	19.	t
10.	a	20.	r

Completion

1. smaller, larger, luminosity
2. surface temperature, hotter, cooler
3. mass, diameter, larger, smaller

Self-Test

(Page numbers refer to the textbook.)

1. d (Objective 1; page 154)
2. c (Objective 1; page 154)
3. d (Objective 2; pages 155–156; video program)
4. b (Objective 2; pages 155–156)
5. a (Objective 3; page 162)
6. a (Objective 3; pages 159–160)
7. b (Objective 4; page 169; video program)
8. a (Objective 4; page 165; video program)
9. d (Objective 5; page 173)
10. c (Objective 5; page 169)

Stellar Formation

Assignments

For the most effective study of this lesson, we suggest that you complete the assignments in the sequence listed below.

Before Viewing the Video Program

- Read the Overview and Learning Objectives for this lesson. Use the Learning Objectives to guide your reading, viewing, and thinking.

- Read Chapter 9, "The Formation and Structure of Stars," section 9–1, pages 179–188, and "Perspective—The Orion Nebula," pages 202–205, in the textbook.

- Read the Viewing Notes in this lesson.

View "Stellar Formation" Video Program

After Viewing the Video Program

- Briefly note your answers to questions listed at the end of the Viewing Notes.

- Review all reading assignments for this lesson, especially the first two paragraphs of the textbook Chapter 9 summary on page 206 and the Viewing Notes in this lesson.

- Write brief answers to review questions 1–4 at the end of the textbook Chapter 9 to be certain you understand the text material.

- Complete the Review Activities in this guide to reinforce your understanding of important terms and concepts. Check your answers with the Answer Key at the end of this lesson and review when necessary.

- Take the Self-Test in this guide to measure your achievement of the Objectives. Check your answers with the Answer Key and review when necessary.

- Complete any other activities and projects assigned by your instructor.

Overview

The previous lesson explored how astronomers are able to learn about stars, in spite of their incredible distances, and what they have learned about them. This lesson examines what astronomers have learned about how stars come to exist. Following study of this first stage of the stellar life cycle, the next three lessons cover the lives of stars, the deaths of stars, and the ultimate fates of the most massive stars.

We know that stars are huge spheres of mostly hydrogen and helium held together by their gravity and shining from the nuclear reactions taking place deep within them. Not surprisingly, stars first form from the interstellar medium—the large, cold, thin clouds of hydrogen and helium found floating throughout our galaxy.

In trying to learn about stellar formation, astronomers originally faced an enormous problem: they literally could not see the places where stars were being born. Astronomers were unable to observe the birth of stars through ordinary telescopes for two principal reasons: (1) the stars are cool when they are first forming and emit very little visible light and (2) stars usually form within thick clouds of dust that absorb what little light they do emit. Because of these two problems, early theories about stellar formation were based mainly on conjecture rather than on specific observations.

The arrival of the space age, however, has given astronomers new tools to use in studying the birth of stars. Sophisticated instruments, such as orbiting infrared telescopes and microwave telescopes, enable astronomers to see through the thick clouds of dust. Now, for the first time, astronomers can witness many of the events that lead to the formation of stars.

Learning Objectives

When you have completed all assignments for this lesson, you should be able to:

1. Describe the composition and distribution of the interstellar medium and how its contents are detected at visible and nonvisible wavelengths. TEXTBOOK PAGES 179–182; VIDEO PROGRAM.

2. Outline the theorized process by which stars form, from molecular cloud to main sequence. TEXTBOOK PAGES 182–185; VIDEO PROGRAM.

3. Describe confirming observations of actual pre-main-sequence objects at visible and nonvisible wavelengths. TEXTBOOK PAGES 185–188 and 202–205; VIDEO PROGRAM.

Viewing Notes

"Stellar Formation" begins with a brief history of what earlier astronomers had theorized about the dark "holes" that seemed to exist between the stars and how these scientists eventually came to learn about the presence of the interstellar medium and about its composition. Several astronomers describe the interstellar medium and explain how its components obscure our observation of stellar nurseries.

The next segment of the video program outlines the conditions necessary for stellar formation and the processes by which stars come into existence. The program pays particular attention to the Orion Nebula, an area of active star formation that is especially easy to study because of its closeness.

The many sophisticated techniques now available to astronomers to see through the interstellar dust are the focus of the concluding segment of the video program. Among the new tools that have provided astronomers with data about stellar formation are the IRAS satellite, which detected infrared light, and the Kuiper Airborne Observatory, which has enabled astronomers make some infrared observations from above most of Earth's atmosphere. The final segment also includes information on T Tauri stars and their associated variable nebulae known as Herbig-Haro objects. When gas jets flow out from these stars, they rapidly clear away the surrounding dust, finally enabling us to see the young stars, new, twinkling points of light.

As you watch the video program, consider the following questions:

1. How did astronomers come to conclude that dark nebulae were not holes in space but clouds of opaque dust?

2. What kinds of gas and dust are found in the interstellar medium? Under what circumstances can the interstellar medium be seen with optical telescopes, or with infrared or radio telescopes?

3. Why do stars form only from cold, dark clouds, and what kinds of external events can trigger such a cloud to form stars?

4. What is the significance of stars forming in clusters, such as in the Orion Nebula, and how can this lead to the formation of binary star systems?

5. What is the purpose of the Kuiper Airborne Observatory?

6. In what ways can we actually observe protostars at different stages of their development? How do T Tauri stars and Herbig-Haro objects relate to star formation?

Review Activities

Matching

Match the terms listed below with the definitions that follow. Check your answers with the Answer Key and review any terms you missed.

_c_ 1. interstellar medium

_f_ 2. interstellar reddening

_h_ 3. infrared cirrus

_e_ 4. evolutionary track

_g_ 5. emission nebula

_a_ 6. reflection nebula

_k_ 7. molecular cloud

_i_ 8. shock wave

_l_ 9. association

_b_ 10. protostar

_d_ 11. nebula

_j_ 12. birth line

_h_ 13. T Tauri star

_m_ 14. Herbig-Haro object

_p_ 15. bipolar flow

_l_ 16. Bok globule

a. any glowing cloud of gas or dust in space reflecting the light of nearby stars

b. a collapsing cloud of gas and dust destined to become a star

c. the gas and dust distributed between the stars

d. a dense interstellar gas cloud in which atoms can link together

e. the path a star follows on the Hertzsprung-Russell (H-R) diagram as it changes its surface temperature and luminosity

f. the process in which interstellar dust scatters away blue starlight

g. a cloud that glows because of starlight reflecting off interstellar dust

h. glowing interstellar gas cloud caused by ultraviolet light from hot stars

i. a sudden change in pressure that travels through space as a sound wave and can trigger star formation

j. the location on the H-R diagram where protostars of different masses first become visible

k. wispy network of cold interstellar dust clouds

l. group of widely scattered stars moving together through space but not gravitationally bound into clusters

m. small, dark cloud of gas and dust believed to be about to form stars

n. young stars that seem to be clearing away surrounding gas and dust

o. the process by which T Tauri stars blow away surrounding material

p. small, variable nebula caused when a T Tauri star excites nearby interstellar gas

Completion

Fill each blank with the most appropriate term from the list for that paragraph. A term may be used once, more than once, or not at all. Check your answers with the Answer Key and review when necessary.

1. Most of the interstellar medium is in the form of __gas__, of which about 75 percent consists of the chemical element __hydrogen__ and 25 percent of the element __helium__. Only about 1 percent of the mass of interstellar material is in the form of solid __dust__. This small amount of solid material creates the phenomenon of interstellar __reddening__, which results from the fact that the longer wavelengths of __red__ photons are __less__ likely to be deflected than the shorter wavelengths of __blue__ photons.

blue	hydrogen
bluing	less
dust	more
gas	red
helium	reddening

2. Stars form from clouds that are __colder__ in temperature and __greater__ in density than normal. As the fragments of such a cloud continue collapsing to form stars, each fragment becomes steadily __hotter__ in temperature and __greater__ in density until nuclear reactions begin and halt the collapse.

colder	hotter
greater	lesser

3. The more massive a protostar, the ___less___ amount of time it takes to form. During most of its existence, a protostar is detectable at ___infrared___ wavelengths, because it is ___colder___ in temperature than normal stars and is also surrounded by ___dust___. The T Tauri phenomenon is a protostar in the process of becoming ___visible___. Herbig-Haro objects are clouds of glowing ___gas___ lying on either side of a protostar detectable at ___infrared___ wavelengths, or on either side of a T Tauri star detectable at ___visible___ wavelengths.

colder	infrared
dust	lesser
gas	nonvisible
greater	ultraviolet
hotter	visible

Self-Test

Select the one best answer for each question. Check your answers with the Answer Key and review when necessary.

1. The element that makes up most of the mass of the interstellar medium is
   a. hydrogen.
   b. helium.
   c. carbon.
   d. silicon.

2. Starlight passing through interstellar dust becomes
   a. fainter and bluer.
   b. brighter and bluer.
   c. fainter and redder.
   d. brighter and redder.

3. The wavelength at which the cold interstellar dust known as "cirrus" can be observed is

   a. gamma ray.

   b. ultraviolet.

   c. visible.

   d. infrared.

4. To begin to collapse into stars, an interstellar cloud must be

   a. cold and thin.

   b. cold and dense.

   c. hot and thin.

   d. hot and dense.

5. The dominant source of shock waves that trigger star formation in our galaxy is thought to be

   a. supernova explosions.

   b. ignition of very hot stars.

   c. collision of two interstellar clouds.

   d. the pattern of spiral arms.

6. A protostar of which of the following masses would take longest to finish forming?

   a. 5 solar masses

   b. 2 solar masses

   c. 1 solar mass

   d. 0.2 solar mass

7. Protostars are observable in the

   a. visible wavelength region.

   b. gamma ray wavelength region.

   c. infrared wavelength region.

   d. ultraviolet wavelength region.

8. Star formation is associated with all of the following objects **except**

   a. red giants.

   b. molecular clouds.

   c. T Tauri stars.

   d. Bok globules.

9. The Herbig-Haro objects associated with any one protostar are frequently seen to occur in groups of

   a. two.

   b. four.

   c. a few dozen.

   d. hundreds.

10. One site of very active star formation is

    a. Crab Nebula.

    b. Orion Nebula.

    c. Cygnus X-1.

    d. Nova Cygni 1975.

Answer Key

Matching

1.	c	9.	l
2.	f	10.	b
3.	k	11.	a
4.	e	12.	j
5.	h	13.	n
6.	g	14.	p
7.	d	15.	o
8.	i	16.	m

Completion

1. gas, hydrogen, helium, dust, reddening, red, less, blue
2. colder, greater, hotter, greater
3. lesser, infrared, colder, dust, visible, gas, infrared, visible

Self-Test

(Page numbers refer to the textbook.)

1. a   (Objective 1; page 180; video program)
2. c   (Objective 1; page 180; video program)
3. d   (Objective 1; page 181)
4. b   (Objective 2; page 182; video program)
5. d   (Objective 2; page 183; video program)
6. d   (Objective 2; page 185)
7. c   (Objective 3; page 185; video program)
8. a   (Objective 3; pages 182, 185, 186)
9. a   (Objective 3; page 186)
10. b   (Objective 3; page 202; video program)

The Lives of Stars

Assignments

For the most effective study of this lesson, we suggest that you complete the assignments in the sequence listed below.

Before Viewing the Video Program

- Read the Overview and Learning Objectives for this lesson. Use the Learning Objectives to guide your reading, viewing, and thinking.

- Read Chapter 9, "The Formation and Structure of Stars," sections 9-2 through 9-4, pages 188–201, and the chapter summary, page 206, in the textbook.

- Read the Viewing Notes in this lesson.

View "The Lives of Stars" Video Program

After Viewing the Video Program

- Briefly note your answers to questions listed at the end of the Viewing Notes.

- Review all reading assignments for this lesson, especially the textbook Chapter 9 summary on page 206 and the Viewing Notes in this lesson.

- Write brief answers to review questions 5–11 at the end of textbook Chapter 9 to be certain you understand the text material.

- Complete the Review Activities in this guide to reinforce your understanding of important terms and concepts. Check your answers with the Answer Key at the end of this lesson and review when necessary.

- Take the Self-Test in this guide to measure your achievement of the Objectives. Check your answers with the Answer Key and review when necessary.

- Complete any other activities and projects assigned by your instructor.

Overview

The last lesson described the formation of a star as a process by which a cloud of interstellar gas collapses under its own weight, becoming denser and hotter at its core until "something" halts the collapse and stabilizes the star. That "something" is thermonuclear fusion, a variety of nuclear reactions capable of generating tremendous energy from the hydrogen that constitutes the bulk of all normal stars. It is when the pressure from these reactions first balances a star's weight that the main sequence phase of any star's life begins—a lengthy phase during which the star changes very little until it eventually runs out of fuel and begins to die.

This lesson is about stars in their main-sequence phase, the most numerous kind, suspended between birth and death. But if previous lessons have made it clear how difficult it is to learn the external properties of stars like luminosity, surface temperature, mass, and diameter, how can we ever learn their internal properties? Basically, we learn about what is hidden deep inside stars through observation and simple physics. The laws of stellar structure are based on common sense and the observation of nature. With these laws and knowledge of a star's external characteristics, computers can create stellar models that detail the star's internal structure at any distance from the center, including whether or not nuclear reactions are occurring in a given layer. A sequence of these models can predict how a star will change as it ages.

These models, painstakingly compared to observations of real stars at various stages of their lives, finally explain why massive stars are more luminous but also more scarce, which was first revealed but left unexplained in Lesson 9.

Learning Objectives

When you have completed all assignments for this lesson, you should be able to:

1. Describe thermonuclear fusion, including the proton-proton chain, CNO cycle, and triple-alpha process and explain how these reactions are governed by a pressure-temperature thermostat. TEXTBOOK PAGES 188–194; VIDEO PROGRAM.

2. Describe the solar neutrino problem along with its possible explanations and implications. TEXTBOOK PAGES 192–193; VIDEO PROGRAM.

3. State the four laws of stellar structure and describe how they can be used to create a model of a star's interior. TEXTBOOK PAGES 195–199.

4. Explain the changes a star undergoes during the main-sequence phase and how the amount of time it spends in that phase is determined. TEXTBOOK PAGES 199–201; VIDEO PROGRAM.

5. Describe the theoretical explanations of the mass-luminosity relation and stellar density function as consequences of the nature of star formation and of the main sequence. TEXTBOOK PAGE 199; VIDEO PROGRAM.

Viewing Notes

The first segment looks at the nuclear reactions themselves, the lifeblood of stars and the process that halts the gravitational collapse that gave birth to the star. The video program concentrates on the proton-proton chain of nuclear reactions, a form of nuclear fusion, which produces most of the energy for the less massive stars such as the sun. Computer animation shows how, in a series of reactions, the hydrogen of a star is transformed into helium and substantial energy is released. In fact, stars generate energy so effortlessly that an energy-hungry humanity dreams of harnessing fusion on Earth, but the segment finishes with a look at the formidable technological difficulties of such a task.

The segment "Stellar Evidence" explains how we know that stars generate nuclear energy deep inside. Astrophysicists describe how computers generate models of a star's interior that, in turn, can predict the star's surface properties. These predictions are validated through comparisons with observations of real stars. The program also looks at how astrophysicists, by studying solar neutrinos, try to gather direct evidence of the sun's reactions. Animation illustrates how neutrinos, particles produced by one of the steps in the proton-proton chain, stream out of the sun instantly instead of taking hundreds of thousands of years to filter to the surface as light does. The segment describes existing and proposed neutrino detectors and highlights the fact that we have detected far fewer neutrinos than we were expecting. Far from being a problem with the sun's nuclear furnace, however, the most likely explanation actually involves the peculiar nature of neutrinos and may affect the future of the entire universe.

The last segment shows how stellar models and theory explain observations of actual stars on the main sequence. Astrophysicists explain how a star's life is a constant battle between pressure and gravity. The evolution of stars is clearly demonstrated on Hertzsprung-Russell (H-R) diagrams. The segment also looks at the hypothesized brown dwarfs, extremely low-mass objects that form like stars, but never become hot enough in their cores to initiate nuclear reactions. Finally, in setting the stage for the next lesson, the video program describes how a star begins to die the moment it finishes being born; the changing chemical composition within even main-sequence stars produces changes that accelerate near the end of a star's life and cause it to swell into a giant—large and bright but with not much time left.

As you watch the video program, consider the following questions:

1. What is the fuel in the proton-proton chain? What kinds of particles are produced? How is the energy generated related to the mass of the helium?

2. What is a stellar model and what does it tell about a star?

3. What are different ways to detect neutrinos? What are some of the explanations for the shortage of neutrinos detected from the sun?

4. Why are main-sequence stars so stable? Why are the most massive stars the most luminous and the shortest-lived?

5. What happens when a main-sequence star runs out of hydrogen in its core? How do observations of star clusters verify this conclusion?

Review Activities

Matching

Match the terms listed below with the definitions that follow. Check your answers with the Answer Key and review any terms you missed.

_____ 1. strong force

_____ 2. weak force

_____ 3. nuclear fission

_____ 4. Coulomb barrier

_____ 5. proton-proton chain

_____ 6. CNO cycle

_____ 7. neutrino

_____ 8. triple-alpha process

_____ 9. nuclear fusion

_____ 10. continuity of mass

_____ 11. continuity of energy

_____ 12. hydrostatic equilibrium

_____ 13. energy transport

_____ 14. opacity

_____ 15. stellar model

_____ 16. brown dwarf

_____ 17. zero-age main sequence (ZAMS)

a. the main, three-step energy source in the sun

b. the main, carbon catalyst energy source in more massive stars

c. the main energy source when a star becomes a red giant, with helium as the fuel

d. the process by which stars generate energy

e. the nuclear process used in uranium-fueled nuclear power plants

f. repulsion between nuclei overcome by violent collision

g. the reason it is difficult for electromagnetic radiation to escape a star

h. balance between material pressing down on a layer and pressure of gas in that layer

i. process by which hot objects cool

j. the reason there are no gaps within a star's layers

k. the reason a star's luminosity depends on its nuclear reaction rate

l. numerical description of conditions within each of a star's layers

m. massless neutral atomic particle that travels at the speed of light

n. the natural force binding protons and neutrons together

o. the natural force responsible for some forms of radioactive decay

p. a hypothetical star with too low a mass to ignite nuclear reactions

q. the location on an H-R diagram occupied by a star that has just stabilized with hydrogen-burning

Completion

Fill each blank with the most appropriate term from the list for that paragraph. A term may be used once, more than once, or not at all. Check your answers with the Answer Key and review when necessary. If a question requires two or more answers in succession, they may be in any order, unless the question indicates otherwise.

1. In the proton-proton chain and CNO cycle, four __hydrogen__ nuclei fuse to form a single nucleus of __helium__. In the CNO cycle, __carbon__ is needed as a catalyst. In the triple-alpha process, the fuel is __helium__, and the principal byproduct is __carbon__. The proton-proton chain dominates in the __less__ massive main-sequence stars and the CNO cycle in the __more__ massive stars.

carbon   less

helium   more

hydrogen   nitrogen

2. If a star is not generating enough energy to balance its weight, its diameter will _decrease_, its central temperature will _increase_, and its reaction rate will _increase_, until it is in balance. The more massive a main-sequence star is, the _more_ luminous it will be and the _less_ time it will spend in the main-sequence phase. These massive main-sequence stars are _less_ numerous than other types of stars.

decrease   less

increase   more

3. Because of hydrostatic equilibrium, as you go from the surface of a star to its center, its pressure will _increase_ and its temperature will _increase_. The three ways in which energy flows from hot regions to cooler regions are _convection_, _conduction_, and _radiation_. Close contact between molecules is necessary for _conduction_ to occur. This method of energy transport is unimportant for most stars because their matter is _gaseous_. In the sun's inner regions, the principal means of energy transport is _radiation_. In this type of energy flow, resistance to the flow is a function of the opacity of the gas. Heat builds up when the opacity of the gas is _high_, creating a churning motion known as _convection_. This churning motion is important because it tends to make the gas _uniform_.

conduction   increase

convection   low

decrease   radiation

gaseous   uniform

high   unstable

4. A hypothetical brown dwarf is ___less___ massive than necessary to ignite nuclear reactions; over time, its temperature will _decrease_

decrease   less

increase   more

Self-Test

Select the one best answer for each question. Check your answers with the Answer Key and review when necessary.

1. In a main-sequence star, the number of hydrogen nuclei necessary to form a single helium nucleus via nuclear fusion is

    a. one.

    b. two.

    c. three.

    d. four.

2. The percentage of the mass of hydrogen that is converted to energy in nuclear fusion is

    a. 100 percent.

    b. 99 percent.

    c. 50 percent.

    d. 1 percent.

3. The first neutrino detector to be built contained

    a. heavy water.

    b. chlorine.

    c. helium.

    d. gallium.

4. The number of solar neutrinos detected by scientists so far is about

    a. 100 times as many as expected.

    b. twice as many as expected.

    c. the number expected.

    d. a third as many as expected.

5. If you hold the bowl of a spoon in a flame, the handle grows warm because of

   a. conduction.

   b. radiation.

   c. convection.

6. "The weight on each layer is balanced by the pressure in that layer" is the law of

   a. continuity of mass.

   b. continuity of energy.

   c. hydrostatic equilibrium.

   d. energy transport.

7. During the main-sequence phase, stars gradually become

   a. fainter and cooler.

   b. fainter and hotter.

   c. brighter and hotter.

   d. brighter and cooler.

8. A main-sequence star is twice as massive as the sun and twelve times more luminous. Its life span will be

   a. the same as the sun's.

   b. six times longer than the sun's.

   c. twice as long as the sun's.

   d. six times shorter than the sun's.

9. In comparison with main-sequence stars, brown dwarfs would be

   a. more massive, smaller, and brighter.

   b. more massive, smaller, and fainter.

   c. less massive, larger, and brighter.

   d. less massive, smaller, and fainter.

10. In comparison with high-mass stars, low-mass stars are

   a. more common because they form naturally in larger numbers
      and have longer life spans.

   b. less common because they form naturally in fewer numbers
      and have shorter life spans.

   c. less common, even though they form naturally in larger
      numbers, because they have shorter life spans.

   d. more common, even though they form naturally in fewer
      numbers, because they have longer life spans.

Answer Key

Matching

1.	n	10.	j
2.	o	11.	k
3.	e	12.	h
4.	f	13.	i
5.	a	14.	g
6.	b	15.	l
7.	m	16.	p
8.	c	17.	q
9.	d		

Completion

1. hydrogen, helium, carbon, helium, carbon, less, more

2. decrease, increase, increase, more, less, less

3. increase, increase, conduction, convection, radiation, conduction,
   gaseous, radiation, high, convection, uniform

4. less, decrease

Self-Test

(Page numbers refer to the textbook.)

1. d   (Objective 1; page 190; video program)

2. d   (Objective 1; page 190)

3. b   (Objective 2; page 192; video program)

4. d   (Objective 2; pages 192–193; video program)

5. a   (Objective 3; page 196)

6. c   (Objective 3; page 198)

7. d   (Objective 4; page 200; video program)

8. d   (Objective 4; page 201)

9. d   (Objective 5; page 199; video program)

10. a   (Objective 5; page 201; video program)

The Deaths of Stars

Assignments

For the most effective study of this lesson, we suggest that you complete the assignments in the sequence listed below.

Before Viewing the Video Program

- Read the Overview and Learning Objectives for this lesson. Use the Learning Objectives to guide your reading, viewing, and thinking.

- Read Chapter 10, "The Deaths of Stars," pages 209–233, in the textbook.

- Read the Viewing Notes in this lesson.

View "The Deaths of Stars" Video Program

After Viewing the Video Program

- Briefly note your answers to questions listed at the end of the Viewing Notes.

- Review all reading assignments for this lesson, especially the textbook Chapter 10 summary on page 232 and the Viewing Notes in this lesson.

- Write brief answers to all the review questions at the end of textbook Chapter 10 to be certain you understand the text material.

- Complete the Review Activities in this guide to reinforce your understanding of important terms and concepts. Check your answers with the Answer Key at the end of this lesson and review when necessary.

- Take the Self-Test in this guide to measure your achievement of the Objectives. Check your answers with the Answer Key and review when necessary.

- Complete any other activities and projects assigned by your instructor.

Overview

The preceding lessons described how stars are born and live; this one explains how they die. The process of a star's death is influenced by its mass and by gravity. Stars come in different masses for which the inward force of gravity is correspondingly different.

Gravity is the driving force behind a star's entire life cycle. It is gravity that collapses a cloud to form a star in the first place, that holds the star together while it is shining, and that waits patiently for the star to run out of energy so it can crush the star into a tiny, dense heap. Thus, a star's mass determines how quickly it forms, how quickly and brightly it lives, and how quickly it dies. And it also determines how it dies.

This lesson divides the subject of stellar deaths into three parts, each of which depends on mass. Low-mass red dwarfs collapse directly into white dwarfs, which is described only in the text. The video program looks at the fate of medium-mass stars such as our own sun, which one day will expand into a giant before blowing off its outer layers as a planetary nebula and then collapsing into a white dwarf. The sun's demise (which will, of course, also destroy any life then on Earth) will therefore be gradual. But if a medium-mass star has a close binary companion star, mass transfer can lead to a more violent death in a nova or Type I supernova explosion.

The most massive stars self-destruct, even without the help of companions, in Type II supernovae when their dense iron cores suddenly collapse and then eject the stars' outer layers. This lesson looks at the part of a supernova that is blown outward, enriching the interstellar gases with the heavier chemicals that formed deep inside the star during its life. It is the next lesson, however, that considers what could become of that collapsing iron core.

Learning Objectives

When you have completed all assignments for this lesson, you should be able to:

1. Describe the internal processes that cause an aging main-sequence star to expand into a giant. TEXTBOOK PAGES 210–214; VIDEO PROGRAM.

2. Describe the deaths of low-mass red dwarfs and of medium-mass stars, including the formation of planetary nebulae and white dwarfs. TEXTBOOK PAGES 214–219; VIDEO PROGRAM.

3. Describe the deaths of the most massive stars, including Type II supernovae and supernova remnants. TEXTBOOK PAGES 219–224; VIDEO PROGRAM.

4. Explain how Hertzsprung-Russell (H-R) diagrams for different star clusters can be used to reveal details of the process of stellar evolution. TEXTBOOK PAGES 224–226 and 227–228.

5. Describe the evolution of close binary stars, including possible effects of mass transfer such as novae and Type I supernovae. TEXTBOOK PAGES 224, 226 and 229–232.

Viewing Notes

The "Deaths of Stars" begins by describing the eventual fate of our own sun, which is typical of the fate of most medium-mass stars. The video program outlines the events occurring in each of the major stages: depletion of hydrogen in the core, the subsequent cessation of nuclear reactions in the core, contraction of the helium core, continued nuclear reactions in the layers surrounding the core, expansion of the sun into a red giant, and, ultimately, the shedding of the outermost layers into a bubble of gas known as a planetary nebula (an inappropriate term, since a planetary nebula is not related to planets), with a white dwarf as the only remnant of our sun.

Although the death of Earth's sun will be gradual, other stars have much more sudden deaths, creating the equivalent of stellar fireworks. The video program explains the three major types of explosive events: nova, which occurs when a white dwarf receives hydrogen-rich outer gases from a companion star; Type I supernova, when a white dwarf's carbon interior ignites; and Type II supernova, when the iron core of a very massive star implodes.

The concluding segment of the video program focuses on supernovae that have been observed on Earth. Particular attention is given to SN1987A, a supernova that was first observed on February 24, 1987, and what astronomers and astrophysicists have learned from the data they have gathered and their observations. The video program also discusses the significance of a supernova explosion, which sweeps a star's outer layers into space, distributing heavier chemicals formed by nuclear processes inside the star throughout the galaxy. These chemicals, including carbon, provide the atoms that are the basis for life on Earth.

As you watch the video program, consider the following questions:

1. What is the sequence of events in the death of a star like our sun, and what observable phenomena are associated with these events?

2. What are three different ways a dying star can explode? Which explosions are the most destructive? Which explosions are the least destructive? Which require a binary companion? What is left behind, and what is ejected in each case?

3. Why was supernova 1987A historically significant, and what have we learned from studying it?

Review Activities

Matching

Match the terms listed below with the definitions that follow. Check your answers with the Answer Key and review any terms you missed.

_____ 1. helium flash

_____ 2. degenerate matter

_____ 3. planetary nebula

_____ 4. black dwarf

_____ 5. Chandrasekhar limit

_____ 6. carbon detonation

_____ 7. supernova remnant

_____ 8. synchrotron radiation

_____ 9. turn-off point

_____ 10. Roche surface

_____ 11. Lagrangian point

_____ 12. angular momentum

_____ 13. accretion disk

_____ 14. nova

a. conservation of this causes an object to spin faster as it gets smaller.

b. the maximum mass of a white dwarf

c. in a gas of this type, nearly all of the lower energy levels are occupied

d. what a white dwarf becomes when it cools

e. what forms around a white dwarf as matter falls toward it

f. the outer boundary of the volume of space controlled by a star's gravity in a binary system

g. mass flows through this when passing from one star in a binary system to another

h. an explosion resulting from the transfer of matter from a normal star through an accretion disk onto the surface of a white dwarf

i. explosive ignition that occurs in some giant stars at a temperature of 600,000,000 K

j. what is produced by the expanding gas when destructive carbon detonation occurs

k. an explosive event in the cores of some giant stars that does not result in a supernova

l. produced by rapidly moving electrons spiraling through magnetic fields

m. the point at which a cluster's stars move from the main sequence toward the red-giant region

n. an expanding shell of gas released nonexplosively by a dying star

Completion

Fill each blank with the most appropriate term from the list for that paragraph. A term may be used once, more than once, or not at all. Check your answers with the Answer Key and review when necessary. If a question requires two or more answers in succession, they may be in any order, unless the question indicates otherwise.

1. Stars of different masses die in different ways. A red dwarf will collapse directly into a _white dwarf_. A star like the sun will become first a _giant star_, then eject its outer layers in a _planetary nebula_ and finally collapse into a _white dwarf_. If this dying star receives additional mass from a binary companion, it can experience a nondestructive explosion on its surface called a _nova_ or a destructive explosion at its center called a _supernova_. A very massive star will first become a _super giant_ and then, when its _iron_ core collapses, produce a _supernova_.

brown dwarf

carbon

giant star

iron

nova

planetary nebula

supergiant star

supernova

white dwarf

2. A main-sequence star becomes a _giant star_ when it runs out of _hydrogen_ in its core so that its purely _helium_ core collapses. Eventually, the _helium_ can ignite explosively without destroying the star. These new reactions will create a core made largely of _carbon_. If this _carbon_ core ignites explosively, it could cause the star to become a _supernova_.

carbon

giant star

helium

hydrogen

iron

nova

supernova

3. The age of a star cluster containing massive blue stars must be relatively _young_. The age of a star cluster containing only giant stars and low-mass red stars must be relatively _old_. A star cluster's turn-off point also indicates the cluster's age. The lower the temperature of a cluster's turn-off point, the _older_ the cluster is. The higher the cluster's turn-off point, the _younger_ it is.

old

older

young

younger

Self-Test

Select the one best answer for each question. Check your answers with the Answer Key and review when necessary.

1. A main-sequence star's outer layers first begin to expand rapidly into a giant star while
   a. hydrogen is burning in the core.
   b. helium is burning in the core.
   c. hydrogen is burning in a shell around the core.
   d. helium is burning in a shell around the core.

2. Degenerate matter is responsible for all of the following **except**
   a. a gas where pressure does not depend on temperature.
   b. the ejection of a planetary nebula.
   c. the helium flash.
   d. carbon detonation.

3. Of the following types of stars, the one that has never been and can never be a giant star is a
   a. red dwarf.
   b. white dwarf.
   c. black dwarf.
   d. Type II supernova.

4. The shell of gas ejected slowly by a giant star is known as a
   a. supernova remnant.
   b. nova remnant.
   c. planetary nebula.
   d. reflection nebula.

5. A Type II supernova is triggered by the collapse of a star's
   a. hydrogen core.
   b. helium core.
   c. carbon core.
   d. iron core.

6. The most recent supernova visible on Earth to the naked eye occurred in
   a. 1987.
   b. 1604.
   c. 1572.
   d. 1054.

7. A turn-off point corresponding to which of the following spectral classes would indicate the youngest star cluster?
   a. B
   b. A
   c. F
   d. G

8. A main-sequence star of which of the following colors is most likely to be youngest?
   a. red
   b. orange
   c. yellow
   d. blue

9. An explosive, although nondestructive, event on the surface of a white dwarf that is part of a binary system results in a
   a. planetary nebula.
   b. Type I supernova.
   c. Type II supernova.
   d. nova.

10. In the binary star system known as Algol,
   a. both stars are main-sequence stars.
   b. both stars are giant stars.
   c. only the more massive star is a giant.
   d. only the less massive star is a giant.

Answer Key

Matching

1. k
2. c
3. n
4. d
5. b
6. i
7. j

8. l
9. m
10. f
11. g
12. a
13. e
14. h

Completion

1. white dwarf, giant star, planetary nebula, white dwarf, nova, supernova, supergiant star, iron, supernova
2. giant star, hydrogen, helium, helium, carbon, carbon, supernova
3. young, old, older, younger

Self-Test

(Page numbers refer to the textbook.)

1. c (Objective 1; pages 210–211; video program)
2. b (Objective 1; pages 211–213, 216, 218–219)
3. a (Objective 2; page 214)
4. c (Objective 2; pages 215–217; video program)
5. d (Objective 3; page 220; video program)
6. a (Objective 3; page 221; video program)
7. a (Objective 4; pages 227–228)
8. d (Objective 4; pages 227–228)
9. d (Objective 5; pages 230–231; video program)
10. d (Objective 5; pages 229–230)

Neutron Stars and
Black Holes

Assignments

For the most effective study of this lesson, we suggest that you complete the assignments in the sequence listed below.

Before Viewing the Video Program

- Read the Overview and Learning Objectives for this lesson. Use the Learning Objectives to guide your reading, viewing, and thinking.

- Read Chapter 11, "Neutron Stars and Black Holes," pages 235–251, in the textbook.

- Read the Viewing Notes in this lesson.

View "Neutron Stars and Black Holes" Video Program

After Viewing the Video Program

- Briefly note your answers to questions listed at the end of the Viewing Notes.

- Review all reading assignments for this lesson, especially the textbook Chapter 11 summary on pages 250–251 and the Viewing Notes in this lesson.

- Write brief answers to all the review questions at the end of textbook Chapter 11 to be certain you understand the text material.

- Complete the Review Activities in this guide to reinforce your understanding of important terms and concepts. Check your answers with the Answer Key at the end of this lesson and review when necessary.

- Take the Self-Test in this guide to measure your achievement of the Objectives. Check your answers with the Answer Key and review when necessary.

- Complete any other activities and projects assigned by your instructor.

Overview

This lesson completes the discussion of the lives of the stars. The previous lesson indicated that the final fate of a single low- or medium-mass star is to collapse into a white dwarf and then gradually cool into a black dwarf at some distant, future time. Only if such a star has a nearby binary companion is it sometimes possible to generate an explosive death.

The last lesson also explained that the most massive stars can undergo destructive supernova explosions when their iron cores suddenly collapse. You've already seen that the violently ejected outer layers form beautiful supernova remnants. This lesson completes the story by describing what can happen to the collapsing iron cores. Depending on the core's mass, it can become either a neutron star or a black hole, both of which are smaller, denser, and stranger than a mere white dwarf.

Unlike white dwarfs, which shine and can be observed and measured like any other star, neutron stars and black holes would not shine in the usual sense. Because they usually do not provide visible evidence of their existence, detection of neutron stars and black holes is, at the least, challenging. In the case of neutron stars, the accidental discovery of pulsars in 1967 was the breakthrough. Pulsars, which are observed to emit rapid, regular bursts of radio waves (and in a few cases visible light), were determined to be rapidly spinning neutron stars. That theoretical link was cemented by the discovery of a pulsar in the Crab Nebula, a supernova remnant that resulted from a single event the ancient Chinese just happened, fortuitously, to have recorded in A.D. 1054.

As for black holes, astronomers are not 100 percent certain that they've found even one, although several objects are good candidates. In every case, the black hole is identified by the effects of its tremendous gravity on a companion star or other material nearby. Understand that this lesson concerns "stellar mass" black holes—the kind that form from the death of a single massive star. In upcoming lessons, you will learn about the supermassive black holes that may lurk in the cores of giant galaxies and that may each be as massive as millions of stars combined. Indeed, that is the direction the course takes next. The next lesson, by starting with our own galaxy, the Milky Way, will begin to look at how individual stars are gravitationally joined together into the assemblages known as galaxies.

Learning Objectives

When you have completed all assignments for this lesson, you should be able to:

1. Outline the theoretical formation of neutron stars and describe their properties. TEXTBOOK PAGES 235–236; VIDEO PROGRAM.

2. Recount the discovery of pulsars and their identification as rapidly spinning neutron stars. TEXTBOOK PAGES 236–237; VIDEO PROGRAM.

3. Describe the observed properties of pulsars and binary pulsars. TEXTBOOK PAGES 237–244; VIDEO PROGRAM.

4. Describe the theoretical formation of black holes and their properties. TEXTBOOK PAGES 244–248; VIDEO PROGRAM.

5. Recount attempts to observe black holes indirectly and list examples of black hole candidates. TEXTBOOK PAGES 248–250; VIDEO PROGRAM.

Viewing Notes

The video program begins with the story of how observations made almost a thousand years apart provided the first evidence of the existence of dense objects, such as neutron stars, left behind by supernova explosions. The program describes how the Chinese observed a "guest star" in 1054 and how later telescopic observations of the location shown on the Chinese sky chart showed that the guest star was in an area now occupied by the Crab Nebula. Over time, modern astronomers measured the expansion rate of the Crab Nebula and, through backward extrapolation, identified the nebula as the remnant of the 1054 supernova observed by the Chinese. The video program also briefly reviews the characteristics and formation of neutron stars and explains the concept of the Chandrasekhar mass limit. Even though astronomers had developed theories about the existence of neutron stars, they did not expect ever to have direct evidence of such stars because they were thought to be too small and too faint.

The next segment of the video program describes how radio telescopes accidentally discovered pulsars in 1967 (thus providing the first direct evidence of the existence of neutron stars) and how astronomers developed the hypothesis that pulsars are, in fact, rapidly spinning neutron stars. Computer graphics illustrate how the lighthouse theory provides a model to describe the observed radiation from a pulsar. Also in this segment, one of the first astronomers to locate and photograph an optical pulsar, Joseph S. Miller, now director of the Lick Observatory, describes how he and his colleague made their discovery, and the video program shows images of the pulsar.

Although neutron stars are incredibly dense, there is a limit to their maximum mass. If a collapsed star's iron core exceeds even the maximum mass and density of a neutron star, the result is a black hole, the subject of the remaining half of the video program.

Astrophysicists describe the properties and formation of black holes, and computer animation illustrates the concepts about black holes. Astrophysicists also describe evidence for the existence of black holes, with particular attention to an unusual X-ray source called Cygnus X-1.

The concluding segment of the video program concerns the theoretical properties of black holes that have not been verified by observation. Computer graphics illustrate descriptions of stationary "Schwarzschild" black holes with their singularities and event horizons and the rotating "Kerr" black holes with their additional ergospheres. Astrophysicists discuss the highly unlikely possibility of using a black hole as a timewarp or spacewarp and the realistic conclusion that any ordinary spaceship would be ripped apart before it could get anywhere near such a warp. Finally, the video program mentions the postulated gravity waves (which current technology cannot detect) and how the search to understand black holes may challenge our current laws of physics.

As you watch the video program, consider the following questions:

1. What is the evidence that the Crab Nebula is the result of a supernova?

2. Why is there a maximum mass that a white dwarf can have, and what happens when that mass is exceeded?

3. What evidence do we have that pulsars are rapidly spinning neutron stars?

4. Why is it not surprising that most pulsars are not in supernova remnants and most supernova remnants do not contain pulsars?

5. Why is there a maximum mass that a neutron star can have, and what happens when that mass is exceeded?

6. How can we detect the presence of a black hole?

Review Activities

Matching

Match the terms listed below with the definitions that follow. Check your answers with the Answer Key and review any terms you missed.

_a_ 1. neutron star
_e_ 2. pulsar
_b_ 3. glitch
_f_ 4. lighthouse theory
_d_ 5. superconductor
_g_ 6. millisecond pulsar
_h_ 7. singularity

_c_ 8. black hole
_j_ 9. escape velocity
_i_ 10. event horizon
_k_ 11. Schwarzschild radius
_l_ 12. time dilation
_m_ 13. gravitational red shift

a. liquid neutrons under tremendous pressure are this kind of material

b. a sudden change in the periodic signals emitted by a spinning neutron star

c. what can form from the collapse of the most massive stars

d. what can form from the collapse of a star a little too massive to leave a white dwarf behind

e. the detectable evidence of a rapidly spinning neutron star

f. the proposed explanation of a pulsar

g. neutron star of the very most rapid spin rate

h. the object of zero radius into which all the matter of a black hole is thought to fall

i. the boundary within which no event is visible to an outside observer

j. equals the speed of light at event horizon

k. radius of the event horizon around a black hole

l. the reason an object never appears to cross the event horizon when seen from a distance

m. the reason a normal telescope could not detect a light of any brightness if located very near the event horizon

Completion

Fill each blank with the most appropriate term from the list for that paragraph. A term may be used once, more than once, or not at all. Check your answers with the Answer Key and review when necessary.

1. The initial mass of a star influences its fate. A collapsed star with a maximum mass of about 1.4 solar masses is a _White dwarf_. A collapsed star that has maximum mass of about two to three solar masses is a _neutron star_. A collapsed object more massive than two to three solar masses should become a _black hole_. Of these collapsed stars, the one that may be visible as a pulsar is a _neutron star_, and the one that can be seen directly is a _white dwarf_. The type of collapsed star, the existence of which can only be inferred from its gravitational effect on, for example, the material of a nearby binary companion, is a _black hole_. Of the three types of stars, the only one not typically formed from a collapsing iron core is a _white dwarf_.

    black hole   white dwarf

    neutron star

2. Pulsars were first detected through the emission of _radio waves_. The discovery of pulsars is thought to confirm the existence of _neutron stars_. According to theory, neutron stars are left behind by _supernovae_, and pulsars have, in fact, been detected inside remnants of these explosions. As pulsars age, they are observed to spin gradually _slower_, except after a glitch, when they will temporarily spin _faster_. When a binary companion star adds mass to a pulsar, the pulsar will spin _faster_.

    black holes   radio waves
    faster   slower
    gamma rays   supernovae
    neutron stars   white dwarfs
    nova   X rays

3. Black holes are theorized to form if the core of a star collapses and has more than about __two to three__ solar masses. If you were to approach a black hole, as you neared the event horizon, you would see yourself falling __faster__. Someone watching you from a distance would see you falling __slower__ as you neared the event horizon. Once within the event horizon, your speed would always be __slower__ than the speed you need to escape.

faster	slower
four to five	two to three
one to two	

Self-Test

Select the one best answer for each question. Check your answers with the Answer Key and review when necessary.

1. A typical neutron star is about the same diameter as
   a. the sun.
   b. Earth.
   c. North America.
   d. a city.

2. According to theoretical models, the bulk of a neutron star is
   a. solid.
   b. liquid.
   c. gaseous.

3. Pulsars were first detected at
   a. radio wavelengths.
   b. infrared wavelengths.
   c. visible wavelengths.
   d. ultraviolet wavelengths.

4. The first pulsar to be found within a supernova remnant was inside the

   a. Orion Nebula.

   b. Cygnus Loop.

   c. Crab Nebula.

   d. Ring Nebula.

5. The radiation from a pulsar is believed to come in

   a. a burst from the entire surface each time the star expands to maximum diameter.

   b. a burst from the entire surface at each moment the star is expanding most rapidly.

   c. two continuous beams located at the poles around which the star rotates.

   d. two continuous beams located at the magnetic poles.

6. Millisecond pulsars are believed to be

   a. young pulsars that have always been isolated.

   b. young pulsars that have, or had, close binary companions.

   c. old pulsars that have always been isolated.

   d. old pulsars that have, or had, close binary companions.

7. The Schwarzschild radius for a ten solar mass black hole would be

   a. 0.45 centimeters.

   b. 1.5 kilometers.

   c. 15 kilometers.

   d. 10,000 kilometers.

8. If you replaced the sun with a one solar mass black hole, the gravity holding Earth in orbit would become

   a. infinite.

   b. stronger, and Earth would be pulled in.

   c. no different than it is now.

   d. weaker, and Earth would escape into space.

9. Which of the following objects is a likely candidate for containing a black hole?

    a. Hercules X-1

    b. Cygnus X-1

    c. Crab Nebula

    d. Orion Nebula

10. Black hole candidates are conspicuous by their emission of

    a. gamma rays.

    b. X rays.

    c. ultraviolet light.

    d. infrared light.

Answer Key

Matching

1.	d	8.	c
2.	e	9.	j
3.	b	10.	i
4.	f	11.	k
5.	a	12.	l
6.	g	13.	m
7.	h		

Completion

1. white dwarf, neutron star, black hole, neutron star, white dwarf, black hole, white dwarf

2. radio waves, neutron stars, supernovae, slower, faster, faster

3. two to three, faster, slower, slower

Self-Test

(Page numbers refer to the textbook.)

1. d   (Objective 1; page 236)

2. b   (Objective 1; page 239)

3. a   (Objective 2; page 236; video program)

4. c   (Objective 2; page 237; video program)

5. d   (Objective 3; page 238)

6. d   (Objective 3; pages 241, 244)

7. c   (Objective 4: page 246)

8. c   (Objective 4; page 246)

9. b   (Objective 5; pages 248–249; video program)

10. b   (Objective 5; pages 248–249)

Unit II

Background Notes

Solar Magnetic Cycle

As the sun ebbs and flows through its 11-year cycle of magnetic activity, it is not just the number of sunspots that changes. The size and shape of the corona, the intensity of the solar wind, and the incidence of solar flares also fluctuate. The high-speed electrically charged particles that escape the sun during flares are of particular concern on Earth.

In pretechnological times, the occurrence of solar flares was noticeable only in the beautiful displays of auroras (northern and southern lights) that brightened the polar skies when the solar particles plowed into Earth's magnetic field hours or days after leaving the sun. In the twentieth century, however, the effects of solar flares have direct impact on daily life. Subtle alterations in Earth's magnetic field upset navigational equipment and induce surges in power lines. Changes in the atmosphere disrupt long-distance radio communications. Spacecraft are even more susceptible; the additional radiation near sunspot maximum heats the atmosphere so that it extends farther from Earth and increases the frictional drag on orbiting satellites. Both Skylab and the Solar Maximum Mission experienced orbital decay, reentering and burning up in the atmosphere much sooner than would have happened if the last two sunspot maxima had not been so intense. Since both spacecraft were designed to study the sun, it is ironic that the sun was responsible for their premature demise.

On Earth, our magnetic field protects us from the brunt of these flare particles. But as humans spend more time in space beyond the protection of this field, such as when exploring the moon or Mars (which lack significant magnetism), we will have to issue flare warnings so that astronauts have time to take proper shelter from this harmful radiation. A shelter buried under a few feet of lunar or Martian soil should offer sufficient shielding.

Far from being able to predict flares, we don't even understand the 11-year cycle fully. It is a complete mystery why this cycle sometimes ceases for extended periods. The most famous instance was the Maunder minimum, first recognized by Walter Maunder about a century ago. The Maunder minimum refers to the period from 1645 to

1715, when a scarcity of sunspots, of auroral displays, and of prominent coronas during eclipses all pointed to a prolonged lack of solar activity. These years also marked the peak of the "Little Ice Age" in Europe and America, with its severe winters and short growing seasons.

Even today, scientists are uncertain whether the Maunder minimum and the Little Ice Age were related or merely coincidental. Written sunspot records go back no farther than Galileo's day, so we are unable to learn if there had been similar occurrences in the past. However, measurements of radioactive carbon 14 in tree rings can also indicate the level of solar activity during the year that each ring formed. Such studies have uncovered about a half dozen Maunder-like minima in the last 5,000 years, as well as several sustained maxima. These events correlate well with whatever temperature, winter-severity, and glacial activity data are available. Basically, when the sun is "quiet" for extended periods, Earth gets colder; when the sun is active over a sustained time, Earth get warmer.

This finding implies that the sun is actually brighter when it is covered in spots. It is difficult to measure slight changes in solar brightness from the ground because Earth's own atmosphere varies substantially in transparency over time. But recent observations from space show the sun is indeed a few tenths of a percent brighter when active. In fact, solar activity has been particularly intense during the last half century, so the global warming trend, which has so frequently been blamed on human pollution of the atmosphere with "greenhouse" gases, may actually be, at least to some degree, a byproduct of natural solar variability.

Even if global warming results in part from natural phenomena, it is small consolation. In a world that already has trouble growing enough food for its burgeoning population, another mini-ice age would not be welcome. We have not been observing the sun with precision long enough to understand its true behavior; the 11-year cycles may be mere ripples upon a larger cycle. We have even been studying magnetic behavior in other stars to gain some insight into the fraction of time a star spends with its cycles either stalled or stuck in high gear. But until the results are in, the inconstancy of the sun is yet another variable to be thrown into our attempts to estimate the future of our climate and our species.

The Milky Way

Assignments

For the most effective study of this lesson, we suggest that you complete the assignments in the sequence listed below.

Before Viewing the Video Program

- Read the Overview and Learning Objectives for this lesson. Use the Learning Objectives to guide your reading, viewing, and thinking.

- Read Chapter 12, "The Milky Way," pages 255–280, in the textbook.

- Read the Viewing Notes in this lesson.

View "The Milky Way" Video Program

After Viewing the Video Program

- Briefly note your answers to the questions listed at the end of the Viewing Notes.

- Review all reading assignments for this lesson, especially the textbook Chapter 12 summary on page 278 and the Viewing Notes in this lesson.

- Write brief answers to all the review questions at the end of textbook Chapter 12 to be certain you understand the text material.

- Complete the Review Activities in this guide to reinforce your understanding of important terms and concepts. Check your answers with the Answer Key at the end of this lesson and review when necessary.

- Take the Self-Test in this guide to measure your achievement of the Objectives. Check your answers with the Answer Key and review when necessary.

- Complete any other activities and projects assigned by your instructor.

Overview

Although the "astro" in astronomy means "star," the study of individual stars accounts for only about half of astronomical investigation. Of course, some astronomers join with geologists and meteorologists in the specialized study of the planets of our solar system, while many other astronomers direct their research to the study of stars in groups. Throughout the universe, gravity has bound together individual stars into clusters, galaxies, clusters of galaxies, and even larger structures that were already forming when the universe was very young. In order to understand the history of the universe and the natural forces shaping it, we must first understand these vast stellar systems.

The focus of this lesson is the galaxy of which we are a part: the Milky Way. Our galaxy is a giant, flattened, spinning spiral comprising hundreds of billions of stars. Our solar system orbits slowly in the outskirts of the Milky Way. When, on very dark clear nights, we see the Milky Way as a faintly glowing strip of light across the sky, we have an inside, edge-on view of our own galaxy.

This lesson is a transitional one. It explains the role of the galaxy's spiral structure in triggering the formation of stars from interstellar gas clouds and describes how the heavy chemical content of succeeding generations of stars keeps increasing as a consequence of the supernova outbursts of previous generations. Thus, this lesson fills in the last remaining details of the story of stellar birth and death that began in Lesson 9.

You will also learn about the features of the Milky Way, which will provide a basis for understanding the features of the billions of other galaxies, even though not all of them are flattened and spiral, and some no longer produce stars.

This lesson is the first of five that look not only at galaxies and their sometimes violent interactions, but at the origin, structure, and destiny of the universe itself.

Learning Objectives

When you have completed all assignments for this lesson, you should be able to:

1. Describe the characteristics of open and globular star clusters and explain how the study of their distribution and of variable stars and proper motion enabled Harlow Shapley to estimate the size of the Milky Way and the location of Earth's solar system within it. TEXTBOOK PAGES 256–261; VIDEO PROGRAM.

2. Describe the features of the Milky Way galaxy and of its major components, including the nature of the stellar orbits in each region. TEXTBOOK PAGES 261–264; VIDEO PROGRAM.

3. Explain the reason for differing populations of stars within the galaxy and relate their distribution to the process of the formation of the Milky Way. TEXTBOOK PAGES 264–268; VIDEO PROGRAM.

4. Describe observations of spiral arms within the Milky Way and other galaxies and explain the two major mechanisms that create spiral structure and their role in star formation. TEXTBOOK PAGES 268–273; VIDEO PROGRAM.

5. Describe observations of the nucleus of the Milky Way at nonvisible wavelengths and the nature of the suspected black hole at the center. TEXTBOOK PAGES 274–277; VIDEO PROGRAM.

Viewing Notes

This video program relates the story of how we have learned about the Milky Way and outlines what we now know about our galaxy.

The first segment traces how astronomers have discovered the basic features of the Milky Way. It describes how William Herschel systematically counted the stars and concluded, correctly, that our solar system was within a flattened disk of stars and, incorrectly, that our solar system was in the center. The segment concentrates on the pivotal work of Harlow Shapley. He studied the distribution of globular star clusters, building on the knowledge developed by Henrietta Swan Leavitt of Cepheid variable stars within them, and eventually identified the real center of the Milky Way. Shapley's work showed that our galaxy was much larger than Herschel had thought and that most of it is hidden from the view of ordinary telescopes by interstellar dust. Most importantly, Shapley established the location of our solar system within the Milky Way.

Recognizing that ours is a dense, disk-shaped galaxy with a sparser, spherical halo surrounding it, the next segment considers how the Milky Way may have formed and acquired this particular shape. Animation illustrates the differing orbits of the stars in the halo and the galactic disk. The video program describes Walter Baade's pioneering work when he discovered that the population II stars in the halo, which included the globular clusters, were weak in heavy elements or "metals," but that the population I stars in the disk, including the newly forming young star clusters in the spiral arms, were much richer. Because the metal content of the interstellar medium is continuously enriched by the explosive deaths of old, massive stars, and because the halo contains only stars with very lengthy lifespans, it became clear that population II stars are older than population I stars and that the stars in the halo must have formed earlier than the stars in the disk. The final part of this segment offers the traditional description of how the Milky Way formed from a huge spherical cloud that collapsed into a smaller, flatter, faster-spinning disk. (Lessons 15 and 16 show that this simple scenario was probably modified by several collisions and mergers of smaller gas clouds and galaxies that occurred during the Milky Way's history.)

In the final segment, astronomers and astrophysicists describe what we have learned about the Milky Way's spiral arms and the center of the galaxy. Because the vast amount of interstellar dust makes optical telescopes almost useless in studying the farther regions of the Milky Way, astronomers use radio telescopes to study neutral hydrogen, which radiates at the 21-centimeter radio wavelength. The video program shows some of the "maps" of the Milky Way that have been made with 21-centimeter radio signals and explains how astronomers have been able to trace out the location of hydrogen and its movement within the spiral arms. Computer animation shows how density waves can create and maintain a spiral pattern. Finally, the program looks at phenomena near the center of our galaxy and how they suggest that a massive, dark, and mysterious black hole may be at the very center of the Milky Way.

As you watch the video program, consider the following questions:

1. Why do we see the Milky Way encircling us at night?

2. Why did Herschel conclude that the Milky Way was flattened and our solar system near the center?

3. Why did Shapley conclude that our solar system was not near the center of the Milky Way?

4. What are the major regions, or parts, of the Milky Way? What are the relative ages of stars in each region? How do we know? What does this tell us about the formation of the Milky Way?

5. How do we study the spiral structure of the Milky Way when interstellar dust obscures most of it?

6. What process can create and maintain a spiral pattern, and how does this process relate to star formation?

7. At what wavelengths can we "see" the center of the Milky Way, and what do we find there?

Review Activities

Matching

Match the terms listed below with the definitions that follow. Check your answers with the Answer Key and review any terms you missed.

_____ 1. Magellanic clouds

_____ 2. instability strip

_____ 3. period of a variable star

_____ 4. RR Lyrae variable star

_____ 5. Cepheid variable star

_____ 6. period-luminosity relation

_____ 7. open star cluster

_____ 8. globular star cluster

_____ 9. proper motion

_____ 10. calibration

_____ 11. disk component

_____ 12. kiloparsec (kpc)

_____ 13. spiral arm

_____ 14. spherical component

_____ 15. halo

_____ 16. nuclear bulge

_____ 17. differential rotation

_____ 18. rotation curve

_____ 19. galactic corona

_____ 20. population I and II

_____ 21. metal

_____ 22. spiral tracer

_____ 23. density wave theory

_____ 24. flocculent galaxy

_____ 25. self-sustaining star formation

_____ 26. Sagittarius A

a. the part of the galaxy including the halo and nuclear bulge

b. the part of the galaxy confined to its plane of rotation

c. the spherical cloud of stars at the center of spiral galaxies

d. large spherical region of a galaxy containing globular clusters

e. region of disk containing bright stars, star clusters, gas, and dust

f. an object, such as an O or a B star, used to map spiral arms

g. hypothesized mass outside the halo

h. small group of relatively young stars in the disk

i. large group of old stars in the halo and nuclear bulge

j. small galaxies beyond, but close to, the Milky Way Galaxy

k. region of H-R diagram in which stars pulsate in diameter

l. length of time it takes a variable star to pulsate once

m. variation in the length of the orbital periods of different stars

n. graph of orbital velocity versus radius in the disk of a galaxy

o. intense radio source near the center of the Milky Way

p. explanation of spiral arms as compressions of the interstellar medium

q. explanation of spiral arms as stellar birth triggered by previous stellar birth

r. a type of galaxy with many short spiral segments

s. a comparison of pulsation time and intrinsic brightness

t. establishment of the relationship between two parameters

u. low-mass star that regularly pulsates in less than one day

v. high-mass star that takes more than one day to pulsate regularly

w. rate at which a star moves across the sky

x. a little over 3,000 light-years

y. an atom heavier than helium

z. a stellar distinction based on the abundance of metal

Completion

Fill each blank with the most appropriate term from the list for that paragraph. A term may be used once, more than once, or not at all. Check your answers with the Answer Key and review when necessary.

1. Regarding the age of stars, blue stars tend to be _younger_ than red stars in a region containing only red main-sequence stars and giant stars are probably _older_. Globular clusters contain mostly _older_ stars and open clusters contain relatively _younger_ stars. In a galaxy, the halo and nuclear bulge contain mostly _older_ stars, and the disk and spiral arms contain mostly _younger_ stars. Population I stars are _younger_ and population II stars are _older_. A star that is very metal-poor is _older_.

   older   younger

2. A spiral galaxy with just two well-defined arms is known as a
   grand design galaxy and probably formed as a result of
   density waves compressing the interstellar material to trigger star
   formation. A spiral galaxy with numerous arms, branches, and spurs
   is known as a _flocculent_ galaxy and probably resulted from
   self-s.f regions being wound up by _differential rotation_.

   density waves   grand design

   differential rotation   self-sustaining star
                                    formation
   flocculent

Self-Test

Select the one best answer for each question. Check your answers with the
Answer Key and review when necessary.

1. Compared to open star clusters, globular clusters tend to be

   a. richer in metals and younger.

   b. poorer in metals and younger.

   c. richer in metals and older.

   d. poorer in metals and older.

2. A group of 100 to 1,000 stars in a region 3 to 30 parsecs in
   diameter would be

   a. a galaxy.

   b. an open cluster.

   c. an association.

   d. a globular cluster.

3. Of the following stars, the one that has the longest period of
   variability is

   a. an RR Lyrae star.

   b. the least luminous Cepheid.

   c. the most luminous Cepheid.

4. The approximate diameter of the disk of the Milky Way galaxy is

    a. 1 light-year.

    b. 100 light-years.

    c. 100,000 light-years.

    d. 10 million light-years.

5. In a typical spiral galaxy, stars in the

    a. disk have nearly circular orbits.

    b. disk have highly elliptical orbits.

    c. halo have nearly circular orbits.

    d. galactic plane have highly elliptical orbits.

6. As you go outward from the sun toward the edge of the Milky Way's disk, the orbital velocity of the stars

    a. decreases greatly (Keplerian motion).

    b. decreases slightly.

    c. remains exactly the same.

    d. increases slightly.

7. Which of the following lists of galactic regions is correctly listed in order of increasing metal content of stars commonly found in each region?

    a. disk, halo, spiral arms, nuclear bulge

    b. halo, spiral arms, disk, nuclear bulge

    c. spiral arms, halo, nuclear bulge, disk

    d. halo, nuclear bulge, disk, spiral arms

8. In terms of the observable metal content in the outer layers of stars,

    a. older stars are metal-poor because their nuclear reactions have consumed the metals used in their formation.

    b. older stars are metal-rich because their nuclear reactions formed increasingly heavy elements over time.

    c. younger stars are metal-rich because they formed from gasses enriched by the heavy elements created in the cores of previous stars.

    d. younger stars are metal-poor because it takes much time for heavy elements to condense out of the interstellar gas.

9. In order to trace the spiral structure of the Milky Way over most of its disk, we need to make observations at

   a. gamma ray wavelengths.

   b. X-ray wavelengths.

   c. infrared wavelengths.

   d. radio wavelengths.

10. The number of arms in the Milky Way is

    a. one.

    b. exactly two.

    c. two, but with branches and spurs.

    d. a great many short spiral segments.

11. The galactic center of the Milky Way lies in the constellation

    a. Cygnus.

    b. Orion.

    c. Sagittarius.

    d. Perseus.

12. In terms of mass, the suspected black hole at the center of the Milky Way is

    a. 30 solar masses.

    b. 3,000 solar masses.

    c. 3 million solar masses.

    d. 300 billion solar masses.

Answer Key

Matching

1.	j	10.	t	19.	g
2.	k	11.	b	20.	z
3.	l	12.	x	21.	y
4.	u	13.	e	22.	f
5.	v	14.	a	23.	p
6.	s	15.	d	24.	r
7.	h	16.	c	25.	q
8.	i	17.	m	26.	o
9.	w	18.	n		

Completion

1. younger, older, older, younger, older, younger, younger, older, older
2. grand design, density waves, flocculent, self-sustaining star formation, differential rotation

Self-Test

(Page numbers refer to the textbook.)

1. d (Objective 1; pages 261–262; 265; video program)
2. b (Objective 1; page 258)
3. c (Objective 1; page 258)
4. c (Objective 1; pages 260–261)
5. a (Objective 2; page 263; video program)
6. d (Objective 2; page 264)
7. d (Objective 3; page 265)
8. c (Objective 3; pages 265–266)
9. d (Objective 4; page 269; video program)
10. c (Objective 4; pages 271–273)
11. c (Objective 5; pages 274–276; video program)
12. c (Objective 5; page 276; video program)

Galaxies

Assignments

For the most effective study of this lesson, we suggest that you complete the assignments in the sequence listed below.

Before Viewing The Video Program

- Read the Overview and Learning Objectives for this lesson. Use the Learning Objectives to guide your reading, viewing, and thinking.
- Read Chapter 13, "Galaxies," pages 281–302, in the textbook.
- Read the Viewing Notes in this lesson.

View "Galaxies" Video Program

After Viewing the Video Program

- Briefly note your answers to questions listed at the end of the Viewing Notes.
- Review all reading assignments for this lesson, especially the textbook Chapter 13 summary on pages 300–301 and the Viewing Notes in this lesson.
- Write brief answers to all the review questions at the end of textbook Chapter 13 to be certain you understand the text material.
- Complete the Review Activities in this guide to reinforce your understanding of important terms and concepts. Check your answers with the Answer Key at the end of this lesson and review when necessary.
- Take the Self-Test in this guide to measure your achievement of the Objectives. Check your answers with the Answer Key and review when necessary.
- Complete any other activities and projects assigned by your instructor.

Overview

It was not until the second decade of the twentieth century that astronomers began to develop a reasonably correct understanding of the true size and shape of the Milky Way galaxy and of our solar system's location within it. But many questions about our galaxy remained. For example, was our galaxy the entire universe, or were other galaxies sprinkled throughout an even vaster space? Although astronomers had identified objects they called "spiral nebulae," they did not agree on what these objects were. Some astronomers thought these nebulae were other galaxies, while others thought they were merely spiral clouds of gas within the boundaries of the Milky Way.

In 1920, a debate between Harlow Shapley and Heber Curtis centered on one such object: the Great Spiral Nebula in Andromeda, also known as M 31. Shapley thought it was a cloud of gas within the Milky Way; Curtis thought it was a galaxy outside the Milky Way. The issue was not resolved until 1924, when Edwin Hubble turned the 100-inch Mount Wilson telescope, then the world's largest, in the direction of M 31. He saw that it comprised a huge number of extremely faint stars. The Great Spiral Nebula was indeed a galaxy, as were all the other spiral nebulae at even greater distances.

With Hubble's discovery, the universe had suddenly become a much larger place, and astronomers had an entirely new class of phenomena to study and understand. Hubble began by classifying galaxies according to their shapes; although many were spiral like the Milky Way galaxy, others were elliptical or irregular. Hubble and other astronomers also measured the distance, size, brightness, and mass of individual galaxies, much as astronomers had started to do for stars a century earlier.

But unlike the stars of the Milky Way, which both approach and recede from us as they race our sun around the galaxy's center, the distant galaxies were found to move in a single direction: away from us. Hubble also discovered a simple relationship—the farther a galaxy is from us, the faster it is moving away from us. This Hubble law was the basis of the conclusion that the universe is expanding, possibly as a result of a big bang some 15 billion years ago. Although this lesson describes the formation of galaxies in the early universe and the manner in which they are distributed throughout space, lessons 17 and 18 provide a more complete description of how the distribution originated as well as the many other implications of a big bang.

Learning Objectives

When you have completed all assignments for this lesson, you should be able to:

1. Identify the three major morphological classes of galaxies and describe their characteristics. TEXTBOOK PAGES 282–285; VIDEO PROGRAM.

2. Describe how distance, diameter, luminosity, and mass of a galaxy are measured, and show how these imply the existence of dark matter. TEXTBOOK PAGES 286–292.

3. State the Hubble law and explain the result of Hubble's comparison of the distances and radial velocities of distant galaxies. TEXTBOOK PAGE 289.

4. Describe the factors involved in galactic formation, including the role of collisions between galaxies. TEXTBOOK PAGES 295–299; VIDEO PROGRAM.

5. Describe clusters, superclusters, and other large-scale structures, such as the Great Wall, in the universe and their implications for the process of galactic formation. TEXTBOOK PAGES 293–295 and 299–300; VIDEO PROGRAM.

Viewing Notes

"Galaxies" begins by tracing the story of how astronomers, by studying galaxies, gradually learned that the Milky Way did not comprise the entire universe. This discovery, as described in the video program by Sallie Baliunas, an astronomer at the Smithsonian Astrophysical Observatory, was "the last great step in the Copernican revolution, displacing Earth from the center of the universe, the sun from the center of the universe, and now the galaxy from the center of the universe."

The first segment relates the debate on whether the spiral nebulae astronomers had observed were within or outside the Milky Way. Old film footage and original photographs illustrate the story of how Hubble and his assistant, Milton Humason, identified and measured a Cepheid variable star in the Andromeda galaxy and were able to establish that the distance to that galaxy placed it outside the Milky Way. The video program continues by describing Hubble's ongoing study of galaxies and how, in 1929, his study led to the conclusion that the universe was expanding. His discovery, known as the Hubble law, states the relationship between the distance of a galaxy and the red shift of its spectral lines.

The video program next examines the three principal galactic shapes: spiral, elliptical, and irregular. Photographs of actual galaxies illustrate astronomers' descriptions of the characteristics and features of each shape. Astronomers also discuss current theories of galactic formation, and computer simulations illustrate galactic collisions, star bursts, and mergers. Such interactions are thought to be important influences on the structure and evolution of galaxies.

Gravity, the principal force behind galactic interactions and behind the grand scale of the universe's structure, is explored in the final segment. The video program explains that galaxies exist in clusters bound together gravitationally, such as the Local Group, the galaxy cluster of which the Milky Way is a part. Gravity can cause galaxies within clusters to move toward each other and occasionally collide despite the expansion of the universe. Evidence of such an event has been found within our Local Group, and the video program provides information on possible future collisions of the Milky Way with the Andromeda Galaxy and the Magellanic Clouds. Finally, the program describes the role of gravity in binding clusters into superclusters and superclusters into even larger-scale walls and filaments of galaxies separated by vast voids. Astrophysicists discuss theoretical models of the large-scale structure of the universe and how galaxies stream through space in response to it.

As you watch the video program, consider the following questions:

1. Why was there disagreement over whether other galaxies existed beyond ours, and how was that controversy resolved?

2. From our knowledge of a galaxy's radial velocity (the speed with which it approaches or recedes from us), how can we estimate its distance?

3. What are the characteristics of the different types of galaxies?

4. How do we think galaxies of different shapes formed, and what role did collisions play in that process?

5. What is the significance of the fact that the oldest stars in any type of galaxy are about the same age? What does that say about an evolving universe?

6. How are galaxies distributed throughout the universe? How does this distribution affect the motions of galaxies?

Review Activities

Matching

Match the terms listed below with the definitions that follow. Check your answers
with the Answer Key and review any terms you missed.

_F_ 1. zone of avoidance

_q_ 2. tuning-fork diagram

_V_ 3. elliptical galaxy

_I_ 4. spiral galaxy

_U_ 5. barred spiral galaxy

_S_ 6. irregular galaxy

_K_ 7. megaparsec (Mpc)

_C_ 8. distance indicator

_I_ 9. H II region

_e_ 10. look-back time

_A_ 11. Hubble law

_B_ 12. Hubble constant

_h_ 13. rotation curve method

_I_ 14. cluster method

_I_ 15. velocity dispersion method

_d_ 16. dark matter

_____ 17. rich galaxy cluster

_____ 18. poor galaxy cluster

_M_ 19. ring galaxy

_n_ 20. galactic cannibalism

_S_ 21. supercluster

a. a measure of the rate of expansion of the universe

b. a number that indicates how fast a galaxy at any given distance is moving
   away from us

c. an object of known luminosity or diameter in another galaxy

d. material so far detectable only by its gravitational influence

e. amount by which one sees into the past by observing an object at a given distance

f. region of sky in which almost no external galaxies are visible

g. a cluster of galaxy clusters

h. determination of a galaxy's mass using orbital velocity and orbital radius of stars

i. determination of a galaxy's mass using the range of velocities within a galaxy

j. determination of a galaxy's mass based on the motions of galaxies within a
   cluster

k. a little over 3 million light-years

l. cloud of ionized hydrogen that can be used as a distance indicator

m. believed to result from a head-on collision of two galaxies

n. process by which a large galaxy absorbs a smaller one

o. irregularly shaped group of galaxies, many of them spiral, with no giant ellipticals

p. group of mostly elliptical galaxies spread over a spherical volume about 3 Mpc wide

q. a galaxy-classification system

r. armless, chaotically shaped galaxy with lots of gas and dust, and both young and old stars

s. a round or slightly flattened galaxy with mostly old stars and little gas and dust

t. a highly flattened galaxy with a well-defined disk containing gas, dust, and young stars arranged in curved arcs

u. a galaxy with an elongated nucleus

Completion

Fill each blank with the most appropriate term from the list for that paragraph. A term may be used once, more than once, or not at all. Check your answers with the Answer Key and review when necessary. If a question requires two or more answers in succession, they may be in any order, unless the question indicates otherwise.

1. The type of galaxy with the least dust and gas, and fewest hot stars, is the _elliptical_ galaxy. These galaxies are classified into subgroups, ranging from _E0_ to _E7_. A galaxy with a disk and lots of gas, dust, and star formation is of the _Spiral_ type. Within this classification of galaxies, the one type that has little gas and dust and only a few hot, bright stars is known as an _S0_ galaxy. This type is thought to be intermediate between _Spiral_ and _elliptical_ galaxies. A galaxy with much gas, dust, and star formation but no disk or nucleus is classified as _irregular_. Most of the galaxies that have been catalogued are _Spiral_, but the most common kind of galaxy is believed to be _elliptical_. A rich galaxy cluster would contain almost all _elliptical_ and _S0_ galaxies.

E0	Sa
E7	Sc
elliptical	S0
irregular	spiral

2. Observing an object of known diameter or luminosity in another galaxy can tell us that galaxy's _distance_. This quantity can also be measured by knowing the Hubble constant and the galaxy's _radial vel_. From this information and apparent brightness, we can determine the object's _luminosity_. From this information and angular size, we can determine the object's _diameter_. Observations of the velocities of stars within galaxies or of galaxies within clusters can yield information on a galaxy's _mass_. The conclusion that the vast bulk of matter in most galaxies is dark was reached through comparisons of the _luminosity_ and _mass_ of galaxies.

diameter	mass
distance	radial velocity
luminosity	

3. Once astronomers were able to identify and map distant galaxies, they soon realized that galaxies are not distributed _uniformly_. Rather, galaxies are grouped into _clusters_. A group of galaxies with more than 1,000 galaxies is known as a _rich gal_. Most of the galaxies within this group are _elliptical_, and often have one or more _giant_ galaxies at their centers. Our Milky Way is part of a grouping known as the _local group_, which is an example of a _poor c_. Such a group contains fewer than 1,000 galaxies and its shape is _irregular_. We do not know the exact number of galaxies in this grouping because of obscuring dust clouds in our galaxy known as the _zone of avoidance_

clusters	poor galaxy cluster
dwarf	rich galaxy cluster
elliptical	spiral
giant	uniformly
irregular	zone of avoidance
Local Group	zone of formation
Local Zone	

Self-Test

Select the one best answer for each question. Check your answers with the Answer Key and review when necessary.

1. Of the following types of galaxies, the one that is most circular in apparent shape and featureless is
   a. E0.
   b. E7.
   c. S0.
   d. Sb.

2. Of the following types of galaxies, the one that has the most gas and dust is
   a. E6.
   b. S0.
   c. Sa.
   d. SBc.

3. The type of galaxy that has the greatest range in diameter, mass, and luminosity is
   a. a spiral galaxy.
   b. an elliptical galaxy.
   c. an irregular galaxy.

4. For nearer galaxies, the most accurate distance indicators are
   a. H II regions.
   b. supernovae.
   c. planetary nebulae.
   d. Cepheid variables.

5. If the Hubble constant equals 50 km/sec/Mpc, and a galaxy has a recessional velocity of 5,000 km/sec, then its distance would be
   a. 2.5 Mpc.
   b. 50 Mpc.
   c. 100 Mpc.
   d. 5000 Mpc.

6. Excluding the galaxies in the Local Group, the other galaxies show spectra that are

   a. all blue shifted.

   b. all red shifted.

   c. about equally divided between blue shifted and red shifted.

   d. not shifted at all.

7. Collisions between galaxies

   a. are extremely rare.

   b. occur once in a while, but the stars within them almost never collide.

   c. are frequent, causing most of the stars within them to collide also.

   d. are frequent, but have little effect on either galaxy.

8. Based on our current understanding of galactic formation and evolution,

   a. nearby rich galaxy clusters are thought to consist mainly of elliptical galaxies, although more distant clusters probably have a somewhat larger number of spiral galaxies.

   b. nearby rich galaxy clusters are thought to consist mainly of spiral galaxies, although more distant clusters probably have a somewhat larger number of elliptical galaxies.

   c. rich galaxy clusters contain the same percentage of elliptical galaxies regardless of distance.

   d. rich galaxy clusters contain the same percentage of spiral galaxies regardless of distance.

9. The Local Group contains all of the following galaxies except

   a. the Milky Way.

   b. M 31, the Andromeda Galaxy.

   c. M 51, the Whirlpool Galaxy.

   d. the Large Magellanic Cloud.

10. Of the following groupings of galaxies, the one that is largest in size is the

   a. Local Supercluster.

   b. Local Group.

   c. Great Wall.

   d. Great Attractor.

Answer Key

Matching

1.	f	8.	c	15.	i
2.	q	9.	l	16.	d
3.	s	10.	e	17.	p
4.	t	11.	a	18.	o
5.	u	12.	b	19.	m
6.	r	13.	h	20.	n
7.	k	14.	j	21.	g

Completion

1. elliptical, E0, E7, spiral, S0, elliptical, spiral, irregular, spiral, elliptical, elliptical, S0

2. distance, radial velocity, luminosity, diameter, mass, luminosity, mass

3. uniformly, clusters, rich galaxy cluster, elliptical, giant, Local Group, poor galaxy cluster, irregular, zone of avoidance

Self-Test

(Page numbers refer to the textbook.)

1. a (Objective 1; pages 282–284)

2. d (Objective 1; pages 282–284)

3. b (Objective 2; pages 289–290)

4. d (Objective 2; pages 286–288; video program)

5. c (Objective 3; page 290)

6. b (Objective 3; page 289; video program)

7. b (Objective 4; page 295; video program)

8. a (Objective 4; pages 298–299)

9. c (Objective 5; page 294)

10. c (Objective 5; pages 299–300)

16

Peculiar Galaxies

Assignments

For the most effective study of this lesson, we suggest that you complete the assignments in the sequence listed below.

Before Viewing the Video Program

- Read the Overview and Learning Objectives for this lesson. Use the Learning Objectives to guide your reading, viewing, and thinking.

- Read Chapter 14, "Galaxies with Active Nuclei," pages 303–317, in the textbook.

- Read the Viewing Notes in this lesson.

View "Peculiar Galaxies" Video Program

After Viewing the Video Program

- Briefly note your answers to questions listed at the end of the Viewing Notes.

- Review all reading assignments for this lesson, especially the textbook Chapter 14 summary on pages 316–317 and the Viewing Notes in this lesson.

- Write brief answers to all the review questions at the end of textbook Chapter 14 to be certain you understand the text material.

- Complete the Review Activities in this guide to reinforce your understanding of important terms and concepts. Check your answers with the Answer Key at the end of this lesson and review when necessary.

- Take the Self-Test in this guide to measure your achievement of the Objectives. Check your answers with the Answer Key and review when necessary.

- Complete any other activities and projects assigned by your instructor.

Overview

The last lesson revealed that the universe is an enormous place, populated by billions of galaxies, each with billions of stars, bound together forever by their mutual gravity and shining in unison. Not all galaxies, however, are this simple. This lesson explores "peculiar," or active, galaxies. Some are so bright, compact, and distant that they once challenged some of our most basic conclusions about space and time.

Although these active galaxies are highly luminous, the energy by which astronomers detect their presence does not come entirely from starlight. After these first active galaxies were identified by radio astronomers in the 1950s, astrophysicists observed that they can produce tremendous amounts of energy over many different wavelengths. Further, as the search for active galaxies continued, astronomers found many of these galaxies at even greater distances than the first ones—in the farthest and youngest reaches of the universe.

The first active galaxies detected were radio galaxies. These galaxies release radio energy from two large lobes, or areas of the sky, extending from both sides of the visible galaxy and going far beyond the visible galaxy's borders. Later, in the early 1960s, quasars were discovered. Quasars are so extremely distant that they appear as little more than starlike points of light, but have some of the most unusual properties of all active galaxies.

One of the greatest accomplishments of recent research has been the realization that a single phenomenon, the presence of a supermassive black hole at the center of some galaxies, can explain quasars as well as the many other types of active galaxies. The steady decline in galactic nuclear activity as we go from the distant (and young) quasars to nearer active galaxies and, finally, to the nearly normal, closest (and oldest) galaxies, including our own Milky Way, may simply be the result of a steady reduction in the amount of material available to fall into these black holes and trigger their tremendous energy output. Indeed, as both galaxies and the universe age, they appear to mellow. This type of evolving universe is exactly the kind predicted by the big bang theory, which will be the focus of the next two lessons.

Learning Objectives

When you have completed all assignments for this lesson, you should be able to:

1. Describe observations of active galaxies and the mechanisms proposed to explain their energy output and other characteristics. TEXTBOOK PAGES 303–310; VIDEO PROGRAM.

2. Describe the discovery and characteristics of quasars. TEXTBOOK PAGES 310–315; VIDEO PROGRAM.

3. Explain how quasar distances are estimated and how recent observations contradict some earlier explanations of quasars. TEXTBOOK PAGES 310–315; VIDEO PROGRAM.

4. Describe the current explanation of quasars and their energy sources and the implications of this explanation for the formation of galaxies and early history of the universe. TEXTBOOK PAGES 315–316; VIDEO PROGRAM.

Viewing Notes

In the opening segment of the video program, several astronomers summarize some of the unusual features of these "peculiar" galaxies and the challenges they pose to those who study them.

The second segment, "Shocking Red Shifts," examines the mysteries presented by quasars when they were first discovered in the 1960s. Among those mysteries were the large red shifts in their spectra and the fact that their luminosity was not coming from starlight. This segment explains how astrophysicists have solved some of the problems surrounding quasars and includes footage of Maarten Schmidt, an astrophysicist at California Institute of Technology, describing how he identified highly red-shifted hydrogen Balmer lines in the spectrum of one particular quasar, 3C 273. The even greater red shifts found in other quasars indicated that they were tremendously distant, residing among or beyond the farthest galaxies in the universe, and receding from us at tremendous speeds. Further, the luminosity of such an object at so great a distance implied that a quasar was as luminous as a thousand Milky Ways rolled into one. But astrophysicists also determined that quasars were quite small, with a volume about the size of our solar system, so that their tremendous luminosity had to be generated by an extremely powerful source and one that was luminous at almost all wavelengths. The video program includes images of quasars detected at various wavelengths.

The solution to the major problem raised by quasars—how to explain all that power coming from such a small space—is the topic of the next segment, "A Beacon at the Beginning of Time." Astrophysicists describe the currently accepted model of a quasar—a massive black hole sitting in the center of a normal galaxy—and how this model explains the luminosity of quasars. Computer animation illustrates how a quasar can replenish its fuel through interaction with other galaxies. Astrophysicists also explain the value of quasars in learning about the early universe and the nature of matter and space.

The final segment of the video program focuses on different kinds of active galaxies, such as Seyferts and radio galaxies, which lie at intermediate distances and, thus, are seen at ages intermediate between young quasars and older, nearby, normal galaxies. With lesser gas content and less frequent collisions, these active galaxies show similar but less intense activity than quasars. Computer animation shows how the accretion disk feeds the black hole in an active galaxy and how the disk is surrounded by a torus, a thick ring of dusty material. Astrophysicists describe the problems caused by our viewing angle of individual galaxies and how clever observation techniques can reveal the central core even when hidden by a torus. Like quasars, active galaxies are remnants of cosmic evolution, and studying them helps us understand the history of our universe.

As you watch the video program, consider the following questions:

1. What reasons led astronomers to conclude that quasars are not stars?

2. Why have astronomers concluded that quasars are extremely distant? luminous? small? and young?

3. How does a supermassive black hole explain the properties of quasars?

4. How does a supermassive black hole explain the properties of active galaxies?

5. How does the angle at which we view an active galaxy affect what we see?

6. Why does the intensity of galactic activity change as we look to greater distances?

Review Activities

Matching

Match the terms listed below with the definitions that follow. Check your answers with the Answer Key and review any terms you missed.

_____ 1. radio galaxy

_____ 2. active galaxy

_____ 3. AGN

_____ 4. double-lobed

_____ 5. hot spot

_____ 6. double-exhaust model

_____ 7. head-tail radio galaxy

_____ 8. Seyfert galaxy

_____ 9. BL Lac object

_____ 10. quasar

_____ 11. local hypothesis

_____ 12. relativistic red shift

_____ 13. gravitational lens

_____ 14. Einstein ring

a. small, powerful source of energy believed to be the active core of a very distant galaxy

b. the change in wavelength in the light from an object receding at a speed near that of light

c. a spiral galaxy with an unusually bright, small, active core

d. center of active galaxy that is emitting large amounts of excess energy

e. a galaxy emitting substantial unusual radiation; often in the form of radio waves

f. active core of an elliptical galaxy

g. a galaxy emitting substantial energy at radio wavelengths

h. a galaxy in which radio waves are emitted from two large regions on opposite sides of the galaxy

i. the proposed explanation for a double-lobed galaxy

j. the theory that quasars are not as distant as their red shifts imply

k. an area of especially intense emission in a double-lobed galaxy

l. the presumed result of intergalactic wind on a double-lobed galaxy

m. an effect caused by a nearer galaxy's gravity on the image of a quasar

n. a specific example of a gravitational lens requiring the most precise alignment

Completion

Fill each blank with the most appropriate term from the list for that paragraph. A term may be used once, more than once, or not at all. Check your answers with the Answer Key and review when necessary.

1. The spectral lines of quasars tend to have very _large_ shifts, and these shifts are always toward the _Red_ end of the spectrum. From this shift, astronomers have concluded quasars are very _distant_, that their luminosity is very _high_, and that their energy comes from very _small_ regions, because their energy varies over very _small_ intervals of time. At a quasar's center, astronomers theorize that there is a _black hole_. The mass of this object at the center is very _large_ and its mass is very _earlier_. In comparison to other galaxies, quasars are thought to have formed much _earlier_ and quasars are _more_ likely to interact with other galaxies.

black hole	low
blue	more
distant	near
earlier	nucleus
high	red
large	small
less	

Self-Test

Select the one best answer for each question. Check your answers with the Answer Key and review when necessary.

1. An eruption in the nucleus of an active galaxy can likely be triggered by

   a. a chain reaction of supernovae.

   b. a collision or tidal interaction with other galaxies.

   c. a big bang.

   d. the collapse of an iron core.

2. The radio emission from radio galaxies typically comes
   a. exclusively from the tiny core of each galaxy.
   b. from the same regions that emit visible light.
   c. from two large regions beyond and on opposite sides of the galaxy.
   d. uniformly from a spherical halo surrounding the galaxy.

3. Of the following objects, the one that **must** be a spiral galaxy is a
   a. quasar.
   b. radio galaxy.
   c. BL Lac object.
   d. Seyfert.

4. Most quasars have all of the following characteristics **except**
   a. a strong radio source.
   b. a large red shift.
   c. a starlike appearance.
   d. superluminosity.

5. Most of the lines in the spectrum of the quasar 3C 273 are caused by
   a. calcium.
   b. carbon.
   c. helium.
   d. hydrogen.

6. The speed of a very high radial velocity, red-shift quasar as determined by the relativistic Doppler formula is
   a. greater than that determined by the classical formula.
   b. exactly the same as that determined by the classical formula.
   c. less than that determined by the classical formula.

7. Quasar distances are estimated by

    a. their observed brightness.

    b. observations of Cepheids within them.

    c. parallax.

    d. their red shifts and the Hubble law.

8. All of the following statements are evidence that quasars are extremely distant **except** for the

    a. observation of a supernova in a quasar's host galaxy.

    b. apparent connection of some quasars to galaxies of lesser red shift.

    c. detection of fuzzy light of similar red shift around some quasars.

    d. gravitational lensing of some quasars by nearer galaxies.

9. Quasars are thought to be galaxies

    a. early in their lives although they formed early in the universe.

    b. early in their lives having formed recently.

    c. at about the same stage of their lives as the Milky Way.

    d. in their final stages of existence, having formed long ago.

    e. in their final stages although they formed recently.

10. Of the following objects, the one that produces the most energy is a

    a. supernova.

    b. Seyfert galaxy.

    c. BL Lac object.

    d. quasar.

Answer Key

Matching

1. g
2. e
3. d
4. h
5. k
6. i
7. l
8. c
9. f
10. a
11. j
12. b
13. m
14. n

Completion

1. large, red, distant, high, small, small, black hole, large, earlier, more

Self-Test

(Page numbers refer to the textbook.)

1. b (Objective 1; pages 307–310)
2. c (Objective 1; pages 304–307; video program)
3. d (Objective 1; pages 307–309)
4. a (Objective 2; page 313; video program)
5. d (Objective 2; pages 310–311; video program)
6. c (Objective 3; page 312)
7. d (Objective 3; page 311; video program)
8. b (Objective 3; pages 313–315)
9. a (Objective 4; page 316; video program)
10. d (Objective 4; page 315)

The Big Bang

Assignments

For the most effective study of this lesson, we suggest that you complete the assignments in the sequence listed below.

Before Viewing The Video Program

- Read the Overview and Learning Objectives for this lesson. Use the Learning Objectives to guide your reading, viewing, and thinking.
- Read Chapter 15, "Cosmology," pages 319–341, in the textbook.
- Read the Viewing Notes in this lesson.

View "The Big Bang" Video Program

After Viewing the Video Program

- Briefly note your answers to questions listed at the end of the Viewing Notes.
- Review all reading assignments for this lesson, especially the textbook Chapter 15 summary on pages 339–340 and the Viewing Notes in this lesson.
- Write brief answers to review questions 1 through 5 at the end of textbook Chapter 15 to be certain you understand the text material.
- Complete the Review Activities in this guide to reinforce your understanding of important terms and concepts. Check your answers with the Answer Key at the end of this lesson and review when necessary.
- Take the Self-Test in this guide to measure your achievement of the Objectives. Check your answers with the Answer Key and review when necessary.
- Complete any other activities and projects assigned by your instructor.

Overview

The lessons of this course have been drawing us ever farther from Earth and deeper into the past. With this lesson and the next, we have finally arrived at the ultimate destination: the edge of the observable universe and at the very beginning of time. The subject of these lessons is cosmology, the branch of astronomy concerned with the universe as a whole, without regard for such comparatively small objects as stars and planets and people. Instead, cosmology pursues answers to the grandest questions about the origin, evolution, structure, and fate of the cosmos. In the very truest sense, cosmologists study the big picture.

For the longest time, under the scrutiny of astronomers, the universe seemed eternal and changeless. Newborn stars might condense out of interstellar material even as other, older stars were fading or blowing away. But on a larger scale, not much seemed to vary, particularly when compared to the minute lifetimes of individual astronomers or of modern astronomy itself. Even when galaxies were discovered at the start of the 1920s, astronomers presumed they came and went endlessly like the stars within them. The dominant assumption was that the universe had neither a beginning nor an end.

That assumption was overturned in 1929. In that year, Edwin Hubble discovered the relationship between galactic red shifts and distances, which was described briefly in Lesson 15. From that seemingly obscure observation, astronomers suddenly had evidence that the universe was expanding, that it had a beginning only several times older than the age of Earth, and above all that it was evolving. That means the early universe was very different from today's universe.

The search for clues about the early universe is the focus of cosmology. Cosmology is such a vast subject, and such an active area of current research, that it is covered in two lessons. Briefly, Lesson 17 explains what cosmologists have already learned about the early history of the universe, and Lesson 18 examines areas that cosmologists are currently exploring. Note, however, that the concepts in these two lessons are deeply intertwined.

Learning Objectives

When you have completed all assignments for this lesson, you should be able to:

1. Define Olber's paradox and describe its resolution. TEXTBOOK PAGES 320–321; VIDEO PROGRAM.

2. Identify and define the three basic assumptions of cosmology and describe the effect of the resolution of Olber's paradox on those assumptions. TEXTBOOK PAGES 321–322.

3. Explain how the Hubble law and general theory of relativity predict the expansion of the universe and the geometry of space-time. TEXTBOOK PAGES 322–326; VIDEO PROGRAM.

4. Explain the discovery and latest observations of primordial background radiation and show how these findings support the big bang theory and contradict the steady state theory. TEXTBOOK PAGES 326–330; VIDEO PROGRAM.

5. Recount currently accepted theories about the early history of the universe from the big bang until the beginning of the formation of galaxies. TEXTBOOK PAGES 326–330; VIDEO PROGRAM.

Viewing Notes

"The Big Bang" video program traces the story of how astronomers, like "cosmic archaeologists," have pieced together the story of the birth and evolution of the universe.

The video program begins by considering Olber's paradox—Why is the sky dark at night?—and examining the principal assumptions on which it is based. Michael Seeds, author of the *Horizons* textbook, answers the question and explains how the answer tells us that universe is not eternal, which overturns previous assumptions. The segment continues by reviewing how Hubble and Humason found evidence that the universe is expanding. Several astronomers describe the universe's expansion, with graphic examples and demonstrations.

Once the expansion of the universe was established and accepted by scientists, the search for an explanation of the phenomenon began. The segment "Eternal Questions" looks at the big bang theory and how it eventually gained acceptance over the steady state theory. One of the most important findings that led to the abandonment of the steady state theory was the discovery of the cosmic microwave background radiation, which is the topic of the next segment, "Echoes of Creation."

That segment incorporates historical film footage of Arno Penzias and Robert Wilson, two physicists at the Bell Laboratories, who discovered cosmic background radiation (CBR) in 1964. Astronomers discuss the importance of that discovery and how cosmic background radiation contributes to understanding the history of the universe. This segment also covers the COBE (Cosmic Background Explorer) satellite and the data it has gathered that support the big bang theory.

The concluding segment features astronomers describing current theories about events at the beginning of the universe, starting within the first microseconds and continues through about the first million years. Computer graphics illustrate these descriptions.

As you watch the video program, consider the following questions:

1. Why *is* the sky dark at night?

2. What reasons do we have to think that the universe is expanding?

3. What reasons do we have to think that there was a big bang?

4. What is believed to have taken place in the million years following the big bang?

5. What is the origin of the cosmic microwave background radiation? Why does it correspond to a temperature of less than 3 Kelvin?

6. How has information obtained from the COBE satellite contributed to cosmological research?

Review Activities

Matching

Match the terms listed below with the definitions that follow. Check your answers with the Answer Key and review any terms you missed.

_C_ 1. cosmology

_a_ 2. Olbers' paradox

_d_ 3. homogeneity

_k_ 4. isotropy

_m_ 5. universality

_d_ 6. cosmological principle

_j_ 7. closed universe

_i_ 8. flat universe

_h_ 9. open universe

_B_ 10. big bang theory

_e_ 11. primordial background radiation

_f_ 12. steady state theory

_g_ 13. recombination

a. theory that the universe does not evolve

b. theory that the universe is expanding from a very hot, dense state

c. study of the nature, origin, and structure of the universe

d. assumption that any observer in any galaxy sees the same general features in the universe

e. present appearance, at 2.7 K, of the hot clouds of the big bang

f. stage, within 1 million years or less of the big bang, at which the material in the universe became transparent to radiation

g. apparent conflict about why the sky is dark at night

h. situation in which the average density of the universe is less than that necessary to halt its expansion

i. situation in which the space-time of the universe is not curved

j. situation in which the expansion of the universe will halt

k. assumption that the general features of the universe look the same in every direction

l. assumption that, on the largest scale, matter is uniformly spread throughout the universe

m. assumption that physical laws on Earth apply everywhere

Completion

Fill each blank with the most appropriate term from the list for that paragraph. A term may be used once, more than once, or not at all. Check your answers with the Answer Key and review when necessary. If a question requires two or more answers in succession, they may be in any order, unless the question indicates otherwise.

1. The cosmological principle is the assumption that the features seen by any observer, anywhere in the universe, appear _the same_. This principle is based on two basic assumptions. One assumption, that the average density of matter in space, when viewed on the _largest_ scale, is the same everywhere, is _homogeneity_. The other assumption, that the universe on the _largest_ scale, looks _the same_ in every direction is known as _isotropy_. Another assumption in cosmology is that the laws of nature are _the same_ throughout the _universe_, though not necessarily at all times. This assumption is known as _universality_. The assumptions of the cosmological principle are consistent with the theory that the universe began in a high-temperature, high-density state, known as the _big bang_ theory.

big bang	smallest
different	steady state
galaxy	the same
homogeneity	universality
isotropy	universe
largest	

2. A closed universe is described as _bounded_ and _finite_. The curvature of a closed universe would be _positive_, and the number of distant galaxies would be too _few_. An open universe is described as _infinite_ and _unbounded_. The curvature of an open universe would be _negative_, and the number of distant galaxies would be too _many_. A flat universe is described as _infinite_ and _unbounded_. The curvature

of a flat universe would be _zero_ , and the distribution of galaxies would be _uniform_ . Regardless of the type of universe, no universe has an _edge_ or a _center_.

bounded	many
center	negative
closed	open
edge	positive
few	unbounded
finite	uniform
flat	zero
infinite	

Self-Test

Select the one best answer for each question. Check your answers with the Answer Key and review when necessary.

1. The main reason the night sky is dark is that the universe is
    a. static.
    b. infinite in size.
    c. finite in age.
    d. expanding.

2. Based on Olbers' original assumptions, the night sky should have been
    a. dark.
    b. infinitely bright.
    c. as bright as the surface of an average star.
    d. about 90 percent the brightness of an average star.

3. Isotropy is the assumption that
    a. matter is uniformly spread throughout space.
    b. the universe looks the same in every direction.
    c. the physical laws we know on Earth apply everywhere in the universe.
    d. the universe is expanding.

4. The cosmological assumption that is most likely to fail when astronomers look to very great distances is that of

   a. homogeneity.

   b. isotropy.

   c. universality.

   d. heterogeneity.

5. According to the Hubble law, if Galaxy A is 10 times more distant than Galaxy B, then Galaxy A is

   a. receding 10 times faster than Galaxy B.

   b. approaching 10 times faster than Galaxy B.

   c. receding 100 times faster than Galaxy B.

   d. approaching 100 times faster than Galaxy B.

6. In which kind of universe would the area of even a very large circle be exactly equal to $\pi r2$ (pi times the square of the circle's radius)?

   a. in every kind of universe

   b. closed universe

   c. open universe

   d. flat universe

7. The primordial background radiation is observable primarily at

   a. radio wavelengths.

   b. ultraviolet wavelengths.

   c. X-ray wavelengths.

   d. gamma ray wavelengths.

8. Detection of primordial background radiation

   a. supports the big bang theory.

   b. supports the steady state theory.

   c. contradicts the big bang and steady state theories.

   d. contradicts the big bang theory.

9. The photons we see as the primordial background radiation were emitted approximately

   a. 0.0001 seconds after the big bang.

   b. 4 seconds after the big bang.

   c. 3 minutes after the big bang.

   d. 1 million years after the big bang.

10. Thirty minutes after the big bang, the second most abundant chemical element in the universe, after hydrogen, was

    a. lithium.

    b. helium.

    c. carbon.

    d. beryllium.

Answer Key

Matching

1.	c	8.	i
2.	g	9.	h
3.	l	10.	b
4.	k	11.	e
5.	m	12.	a
6.	d	13.	f
7.	j		

Completion

1. the same, largest, homogeneity, largest, the same, isotropy, the same, universe, universality, big bang

2. finite, bounded, positive, few, infinite, unbounded, negative, many, infinite, unbounded, zero, uniform, edge, center

Self-Test

(Page numbers refer to the textbook.)

1. c   (Objective 1, pages 320–321; video program)
2. c   (Objective 1, page 320; video program)
3. b   (Objective 2, page 321)
4. c   (Objective 2, pages 321–322)
5. a   (Objective 3, page 323; video program)
6. d   (Objective 3, page 325)
7. a   (Objective 4, page 328; video program)
8. a   (Objective 4, page 328; video program)
9. d   (Objective 5, page 330)
10. b   (Objective 5, page 330)

The Fate of the Universe

Assignments

For the most effective study of this lesson, we suggest that you complete the assignments in the sequence listed below.

Before Viewing the Video Program

- Read the Overview and Learning Objectives for this lesson. Use the Learning Objectives to guide your reading, viewing, and thinking.

- Review Chapter 15, "Cosmology," pages 319–341, in the textbook, paying particular attention to "The End of the Universe: A Question of Density," (pages 330–333) and section 15-3, "Refining the Big Bang: Three Questions," (pages 333–339).

- Read the Viewing Notes in this lesson.

View "The Fate of the Universe" Video Program

After Viewing the Video Program

- Briefly note your answers to the questions listed at the end of the Viewing Notes.

- Review all reading assignments for this lesson, especially the textbook Chapter 15 summary on pages 339–340 and the Viewing Notes in this lesson.

- Write brief answers to review questions 6-10 at the end of textbook Chapter 15 to be certain you understand the text material.

- Complete the Review Activities in this guide to reinforce your understanding of important terms and concepts. Check your answers with the Answer Key at the end of this lesson and review when necessary.

- Take the Self-Test in this guide to measure your achievement of the Objectives. Check your answers with the Answer Key at the end of this lesson and review when necessary.

- Complete any other activities and projects assigned by your instructor.

Overview

The last lesson described the universe in its infancy, a time when there was little more than hydrogen and helium. The universe was slowly drawing itself together with gravity to form the young galaxies. The universe was not yet filled with the stars and, presumably, planets and life-forms that tempt us to explore it today. But as much as we've learned about those earliest moments, some very fundamental questions remain to be answered. Although we don't have the answers yet, we will, once we make the necessary observations. In making these observations, we are thwarted mostly by the fact that we must study objects at very great distances in order to see them as they were in the distant past.

For example, astronomers have yet to discern why the space-time of the universe is so flat, and why the initial distribution of material within it is so smooth. The answer probably lies in events that took place less than a second after the beginning of the universe, when the whole universe played by the rules of quantum theory that, today, govern only the tiny spaces inside of atoms.

Astronomers don't yet know how this smooth gas organized itself into the frothy large-scale structure that exists today. That process seems to have happened all too quickly, but that perception is complicated by the fact that we don't know the age of the universe with certainty, nor do we know the nature of the dark matter that seems to make up 90 to 99 percent of the universe and contributed to the organizational process in unknown ways.

Until cosmologists understand that dark matter, we probably will not be able to determine the fate of the universe, for density equals destiny. The amount of material and consequent gravity determine whether the expansion will halt or persist forever. If the universe stops expanding, it may collapse, perhaps triggering another big bang in an endless series of recycling universes. If the universe continues expanding, then the universe happened only once. It may exist forevermore, but the consumption of usable energy would leave it eventually a stagnant place. We are fortunate to inhabit the universe while it is young.

Learning Objectives

When you have completed all assignments for this lesson, you should be able to:

1. Identify unsolved problems in cosmology and describe recent attempts to refine the big bang theory with quantum concepts such as an inflationary universe and Grand Unified Theories (GUTs). TEXTBOOK PAGES 330–335; VIDEO PROGRAM.

2. Briefly relate several possible courses for the future of the universe and the kinds of observations that would be necessary to resolve the issue. TEXTBOOK PAGES 330–333; VIDEO PROGRAM.

3. Describe recent attempts to determine the age of the universe and the contradictions encountered between different techniques. TEXTBOOK PAGES 335–337; VIDEO PROGRAM.

4. Describe recent attempts to establish the existence and nature of dark matter and explain the implications of dark matter for the origin of the universe's large-scale structure and galaxies. TEXTBOOK PAGES 332–333 and 337–339; VIDEO PROGRAM.

Viewing Notes

As a companion program to "The Big Bang," "The Fate of the Universe" completes the story of what cosmologists know today about the evolution of the universe.

Cosmological research is one of the most active areas of astronomical research, and the video program examines several major discoveries that have filled gaps in our knowledge of the evolution of the universe.

This video program includes information on current research in cosmology and details the important contributions of the repaired Hubble Space Telescope (HST). No other telescope can see objects as far or as faint or as clearly.

Of particular interest to cosmologists is the nature of dark matter, which may account for as much as 99 percent of the universe. The program shows how astronomers use gravitational lenses and other techniques to locate this matter and determine the quantity. Understanding of dark matter is the key to understanding the large-scale structure of the universe and the outcome of the universe's expansion.

Another challenge for cosmologists is determining the exact age of the universe. To do so, cosmologists must first learn the precise expansion rate of the universe (the Hubble constant). The video program explores attempts to pinpoint the Hubble constant and concludes with astronomers describing several scenarios for the future of the universe.

As you watch the video program, consider the following questions:

1. How is the Hubble Space Telescope used in cosmological research?

2. How do we go about determining the age of the universe?

3. What reasons do we have to conclude that the universe contains a substantial amount of dark matter?

4. What are some of the forms dark matter may take?

5. How are gravitational lenses and other techniques used to study dark matter?

6. What is meant by an inflationary universe?

7. What factors determine the fate of the universe, and what observations must we make to learn it?

Review Activities

Matching

Match the terms listed below with the definitions that follow. Check your answers with the Answer Key and review any terms you missed.

__c__ 1. critical density

__e__ 2. oscillating universe theory

__b__ 3. horizon problem

__h__ 4. Hubble time

__d__ 5. inflationary universe

__a__ 6. flatness problem

__g__ 7. Grand Unified Theories (GUTs)

__f__ 8. galaxy seeds

a. the peculiar circumstance that the early universe appears to have had exactly the right amount of matter to make space-time flat

b. the circumstance that the primordial background radiation seems more uniform than expected by the big bang theory

c. the quality the distribution of matter in the universe must have in order for space-time to be flat

d. a version of the big bang theory, derived from GUTs, in which the horizon and flatness problems are resolved by a predicted event within the first second after the big bang

e. a situation that is possible only if the universe has greater than critical density

f. possible defects in space-time with strong gravitational fields

g. attempts to describe the electromagnetic, weak, and strong forces of nature in a similar way

h. age of the universe if it had expanded since the big bang at a constant rate

Completion

Fill each blank with the most appropriate term from the list for that paragraph. A term may be used once, more than once, or not at all. Check your answers with the Answer Key and review when necessary.

1. If the density of the universe is exactly equal to the critical density, then the universe is said to be _flat_. If the universe has less than the critical density and expands forever, the universe is said to be _open_, but if the expansion halts at a finite future time, the universe is said to be _closed_. If the universe collapsed and experienced another big bang, it would be described as _oscillating_ If the universe experienced an extremely brief period of extremely rapid expansion immediately after the big bang, then it would be described as _inflational_ Such a brief, rapid expansion is theorized to result from the _strong force_ separating from the _weak force_ and the _____.

closed

electromagnetic force

flat

gravitational force

inflationary

open

oscillating

strong force

weak force

Self-Test

Select the one best answer for each question. Check your answers with the Answer Key and review when necessary.

1. Of the following combinations of types of universes, the one that is **not** possible is the
   a. closed, inflationary, and oscillating universe.
   b. open, inflationary, and oscillating universe.
   c. flat and inflationary universe.
   d. closed and oscillating universe.

2. The force that is **not** unified with the others in a GUT is the
   a. strong force.
   b. electromagnetic force.
   c. weak force.
   d. gravitational force.

3. The inflationary universe theory predicts that the universe must be
   a. completely flat.
   b. nearly flat, but could be barely open or closed.
   c. very open.
   d. very closed.

4. The type of universe in which the average density is predicted to be greatest is the
   a. open universe.
   b. flat universe.
   c. closed universe.

5. If the Hubble constant equals 100 km/sec/Mpc, then the Hubble time equals
   a. 100 billion years.
   b. 20 billion years.
   c. 10 billion years.
   d. 1 billion years.

6. The rate of expansion of the universe in the past is believed to have been

   a. greater than the rate of expansion today.

   b. precisely the same as the rate of expansion today.

   c. less than the rate of expansion today.

7. For any given value of the Hubble constant, the type of universe that would probably be the youngest is the

   a. open universe.

   b. flat universe.

   c. closed universe.

8. The most distant known quasars had formed by the time the universe was no more than

   a. 0.01 percent of its present age.

   b. 1 percent of its present age.

   c. 7 percent of its present age.

   d. 25 percent of its present age.

9. The 0.0006 percent irregularities that COBE found in the cosmic microwave primordial background radiation indicates that

   a. the Milky Way is moving relative to the background.

   b. the radiation is a perfect black-body.

   c. the gas produced by the big bang was perfectly uniform.

   d. large, slightly denser patches of gas had formed within the first million years after the big bang.

10. The percentage of the mass of the universe that is dark matter is estimated to be

    a. 1 to 2 percent.

    b. 10 to 20 percent.

    c. 50 percent.

    d. 90 to 99 percent.

Answer Key

Matching

1.	c	5.	d
2.	e	6.	a
3.	b	7.	g
4.	h	8.	f

Completion

1. flat, open, closed, oscillating, inflationary, strong force, weak force, electromagnetic force

Self-Test

(Page numbers refer to the textbook.)

1. b (Objectives 1 and 2; pages 330–335)

2. d (Objective 1; pages 333–335)

3. b (Objectives 1 and 2; pages 332–334)

4. c (Objective 2; pages 330–332)

5. c (Objective 3; pages 335–336)

6. a (Objective 3; page 335)

7. c (Objectives 3 and 4; pages 335–336)

8. c (Objective 4; page 337)

9. d (Objective 4; pages 338–339)

10. d (Objective 4; page 332)

The Origin of the Solar System

Assignments

For the most effective study of this lesson, we suggest that you complete the assignments in the sequence listed below.

Before Viewing the Video Program

- Read the Overview and Learning Objectives for this lesson. Use the Learning Objectives to guide your reading, viewing, and thinking.

- Read Chapter 16, "The Origin of the Solar System," pages 345–366, in the textbook.

- Read the Viewing Notes in this lesson.

View "The Origin of the Solar System" Video Program

After Viewing the Video Program

- Briefly note your answers to the questions listed at the end of the Viewing Notes.

- Review all reading assignments for this lesson, especially the textbook Chapter 16 summary on pages 363–364 and the Viewing Notes in this lesson.

- Write brief answers to all the review questions at the end of textbook Chapter 16 to be certain you understand the text material.

- Complete the Review Activities in this guide to reinforce your understanding of important terms and concepts. Check your answers with the Answer Key at the end of this lesson and review when necessary.

- Take the Self-Test in this guide to measure your achievement of the Objectives. Check your answers with the Answer Key and review when necessary.

- Complete the Using What You've Learned activities and any other activities and projects assigned by your instructor.

Overview

During the first part of this course, you learned about how astronomers practice astronomy. You learned about the instruments they use and how they analyze light to gain knowledge about the objects producing that light, including our own star, the sun. This knowledge provided a foundation for studying objects that exist far beyond our solar system. You learned about the different stars and the origin, evolution, and death of each type; about black holes; about galaxies and how they cluster; and about quasars. In the last two lessons, you became familiar with theories and observations about the origin and evolution of the universe itself.

The course now returns closer to home and investigates celestial objects that, for many, are more fascinating and exotic than anything else in the universe. These objects are the members of our solar system: the planets, including Earth, and their moons, asteroids, and comets.

Some astronomers and many amateur astronomers find these relatively nearby objects so interesting for several reasons. First, these objects are the only celestial objects whose surfaces we can study in detail. Also, because some of the planets have geologic features not found anywhere on Earth, study of those features can help us to develop theories and models to explain the origin and evolution of these worlds. Second, on some of these other worlds, we can study the remnants of early planetary history, where no such evidence exists on Earth. Third, the planets enable us to study the effects of extreme temperatures, dense and rarefied atmospheres, and large and small gravitational fields. Study of these other worlds enables us to conduct experiments in an immense "laboratory" that cannot be duplicated on the surface of Earth. Finally, study of our solar system allows us to ask and possibly answer some of the most important questions posed by humans: How did Earth form? What influence, if any, did the members of our solar system have on the evolution of Earth and life itself? What can we learn about the future of Earth from studying the environments of the nearby planets and the motions of the smaller members of the solar system? Exploration of our solar system, analysis of the observations, and conclusions drawn from those observations provide us with fascinating views of the planets, their moons, and the other members of the solar system.

This lesson begins the exploration by explaining current knowledge and theory about the origin of the solar system. It covers the characteristics of the solar system and how theoretical models account for the observed characteristics. The video program for this lesson includes planetary scientists describing observations that have lead them to conclude that solar systems beyond our own exist.

Learning Objectives

When you have completed all assignments for this lesson, you should be able to:

1. State several reasons for studying the solar system and cite examples of how knowledge gained from such study has enhanced understanding of the formation of Earth. TEXTBOOK PAGES 345–346, 349–351, and 357–362; VIDEO PROGRAM.

2. List the characteristic properties of the solar system and state current estimates of its age. TEXTBOOK PAGES 346–349 and 356–357; VIDEO PROGRAM.

3. Outline the solar nebula theory and explain how the characteristic properties of the solar system provide evidence that supports that theory. TEXTBOOK PAGES 347 and 357–363; VIDEO PROGRAM.

4. Describe the characteristics of the terrestrial and Jovian planets and describe and compare the processes that resulted in the formation of each. TEXTBOOK PAGES 349–355; VIDEO PROGRAM.

5. Identify the minor members of the solar system and describe their characteristics. TEXTBOOK PAGES 355–356.

6. Outline theories on the formation of planetesimals and the growth of protoplanets. TEXTBOOK PAGES 358–362; VIDEO PROGRAM.

Viewing Notes

"The Origin of the Solar System" opens with computer simulation providing a spectacular flyby of all the planets of our solar system. The video program then moves on to a detailed examination of the solar nebula theory that explains how our solar system was formed. While several astronomers comment on different parts of the theory, computer graphics illustrate the spinning solar nebula, the accretion of dust particles, and the formation of the protoplanetary disk.

The second segment, "Observation," examines how astronomers are searching distant stars to find evidence to support the solar nebula theory. At Kitt Peak, astronomer Frederick M. Walter describes the difficult search for signs of the formation of planets around distant stars in the Great Nebula in Orion, a region of active star formation. The video program also describes the work of radio astronomers in examining disks around new stars. Anneila Sargent, a radio astronomer at Owens Valley Radio Observatory, describes the conclusions she has drawn from her observations of protoplanetary disks, including the details she has seen around HL Tauri.

In searching for evidence of planetary formation, infrared telescopes have been especially valuable in detecting cooler dust disks around older stars. Richard Terrile, a Jet Propulsion Laboratory astronomer, describes the infrared image he captured of Beta Pictoris. That image shows a disk with a hollowed-out center approximately the size of our solar system; Terrile and other astronomers believe planets may be orbiting in the void around the star.

The last part of the video program focuses on the evolution of our solar system, beginning with descriptions of the rocky terrestrial planets and the gas giants. Planetary experts explain the role of heat and gravity in planetary formation and structure. They also describe cratering and explain why cratering is not evident on Earth's surface. Other astronomers explain the role of a planet's distance from the sun in determining its size and atmosphere and describe the concept of the "ice line" in the solar system. The video program also covers the various small bodies of the solar system, including asteroids and moons and concludes with a summary of the solar nebula theory and the implications of that theory for finding similar planetary systems around the other stars in our galaxy.

As you watch the video program, consider the following questions:

1. What is a protoplanetary disk and how is it formed?

2. What processes are involved in planetary formation?

3. What ends the process of planet formation and clears the disk?

4. How have dust disks around other stars been detected and what is the significance of the discovery of these dust disks?

5. What are the principal differences between the terrestrial and Jovian planets of our solar system?

6. What is the process that causes a terrestrial planet to form an iron core and a solid silicate crust?

7. What role did distance and temperature play in making some planets small, dense, and rocky and others large, gaseous, and low density?

8. What is the "ice line" and what role does it play in explaining the formation of the planets?

9. How does understanding the formation of our own solar system help us to understand the formation of solar systems around other stars?

Review Activities

Matching

Match the terms listed below with the definitions that follow. Check your answers with the Answer Key and review any terms you missed.

_e_ 1. solar nebula theory

_m_ 2. Titius-Bode rule

_f_ 3. terrestrial planet

_a_ 4. Jovian planet

_v_ 5. liquid metallic hydrogen

_p_ 6. Galilean satellites

_s_ 7. asteroids

_j_ 8. comet

_i_ 9. dirty snowball theory

_r_ 10. meteor

_c_ 11. meteoroid

_b_ 12. meteorite

_H_ 13. condensation sequence

_L_ 14. planetesimal

_o_ 15. differentiation

_g_ 16. accretion

_y_ 17. protoplanet

_k_ 18. homogeneous accretion

_d_ 19. heat of formation

_n_ 20. radiation pressure

_q_ 21. outgassing

_t_ 22. heterogeneous accretion

a. planet with a large diameter and low density

b. an object that survives its passage through the atmosphere and strikes the ground

c. a rock in space before it enters Earth's atmosphere

d. energy released by in-falling of matter during formation of a planetary body

e. theory that the planets and the sun formed from the same cloud of gas and dust

f. a small, dense, rocky planet

g. the sticking together of solid particles to produce a larger particle

h. order in which different materials condense from a gas in relationship to distance from the sun

i. theory that the nuclei of comets are kilometer-sized chunks of ice with imbedded silicates

j. small, icy body that orbits the sun

k. formation of a planet by the accumulation of planetesimals of the same composition

1. small bodies that formed from the solar nebula and eventually grew into protoplanets

m. simple series of steps that produces numbers approximately matching the sizes of most of the planetary orbits

n. force exerted on the surface of a body by its absorption of light

o. separation of planetary material according to density

p. the four largest satellites of Jupiter

q. creation of a planet's atmosphere from the planet's interior

r. streak of light in the sky

s. minor planets, most of which orbit the sun between Mars and Jupiter

t. formation of a planet by the accumulation of planetesimals of different composition

u. massive object resulting from the coalescence of planetesimals in the solar nebula

v. material that is a good electrical conductor found in the interiors of Jupiter and Saturn

Completion

Fill each blank with the most appropriate term from the list for that paragraph. A term may be used once, more than once, or not at all. Check your answers with the Answer Key and review when necessary. If a question requires two or more answers in succession, they may be in any order, unless the question indicates otherwise.

1. In order to develop a theory that describes the origin of the solar system, the theory must be able to explain the major physical characteristics of the solar system. We can divide these characteristics into two main categories. The first category involves the rotational and orbital characteristics of the sun and planets; the second involves the physical properties of the planets and other minor members of the solar system.

   The sun rotates on its axis in approximately 30 days. If you were looking down on the solar system from the north (above), you would see that the direction in which the sun rotates is _Counter Clockwise_. If you were looking at the motion of the planets, you would observe that the rotation of most of the planets is _Counter Clockwise_. The only exceptions to this observation are the planets _Uranus_ and _Venus_. Furthermore, all the solar system's

__Planets__ and almost all of the _minor members_ orbit the sun in a _counterclockwise_ direction. These observations indicate that the cloud of material from which the solar system originated must have been rotating in the __same__ direction.

The orbits of the planets tend to lie in a _common plane_. Thus, if you view the solar system __edge on__, the planets would appear to be in a __straight__ and __horizontal__ line on either side of the sun. If viewing the solar system from such a perspective, the shape would resemble a __thin disk__. Not only do the __orbits__ of the planets lie within this relatively __thin disk__, but the __equators__ are only __slightly__ tilted with respect to the plane of __Earth's__ orbit.

Based on their physical characteristics, the planets can be divided into __two__ categories. The planets nearest the sun are very similar to planet __Earth__. In size, these planets are relatively __small__, their composition is __solid__, and their surfaces are __rocky__. These planets are also referred to as the __terrestrial__ planets. The size of the planets farthest from the sun is relatively __large__, and their composition is __gaseous__. Also, we are unable to see __solid__ surfaces. These planets appear to be very similar to the planet __Jupiter__ and therefore are referred to as the __Jovian__ planets. Their atmospheres contain the lightest gases, __hydrogen__ and __helium__.

Although the sun and planets are the major components of the solar system, any theory about the origin of the solar system must also account for the system's many minor members. These minor members include __moons__, in orbit around the major planets, __asteroids__, and __comets__. These minor objects are composed of various materials, including __rocks__, __ice__, __gas__ and __dust__. In fact, "dirty snowball" is a term sometimes used for __comets__. An additional and interesting observation of our solar system reveals that it is relatively empty of __gas__ and __dust__, which are remnants of the original cloud of material that formed the solar system. Taking all of these observations into account, astronomers, at present, think that the best explanation of the formation of our solar system is the __nebula theory__.

asteroids	helium	rocky
clockwise	horizontal	same
comets	hydrogen	slightly
common plane	ice	small
counterclockwise	Jovian	solar nebula theory
disk	Jupiter	solid
dust	large	solid and rocky
Earth	minor members	straight
Earth's	moons	terrestrial
edge on	opposite	thin disk
equators	orbits	two
four	oxygen	Uranus
gas	planets	Venus
gaseous	rocks	

2. Planets begin to form when _Planet Plismals_ join to make larger objects called _Proto Planets_. These objects grow large when they acquire additional material that collides with them. These objects do not shatter when they collide with this material because they are traveling in the ___Same___ direction. Therefore, the relative _Velocity_ of the colliding objects are the ___low___ and the materials stick to the larger object's surface. Adherence to the surface is also facilitated because of _sticky coats_ and _sheer forces_ on the surfaces of the smaller objects. Also, the adherence of larger _Particals_ to these objects may be aided by ___gravity___, which helps to attract and hold additional material.

At present, planetary astronomers are debating two theories on the formation of the Earthlike or _terrestrial_ planets. One theory describes how the protoplanets formed by the accretion of _Planetisimals_ that were of the same general _composition_. This theory is called the _homo Accretion_ theory. It assumes that the planets formed ___Slowly___, which allowed the ___Heat___ of formation time to radiate into ___Space___. Once the planet formed, it consisted of matter that was of _similar_ density and uniform _Composition_. Within the interior, _Radioactive_ elements began to decay, generating ___Heat___, which caused the interior of the planet to ___Melt___. Once the interior of the planet turned molten, heavy metals such as ___Iron___ and ___Nickel___ slowly ___Sank___ to the ___Core___ of the planet. The lighter _Silicates_ rose to the ___Surface___, cooled, and formed a _____low density_____ crust. During this process, the original atmosphere probably _____evaporated_____. New gases then _____outgassed_____ from the _____interior_____ of the planet, which then accumulated to form a secondary _____atmosphere_____

The second theory about the formation of the _____system_____ planets is described as the _____Hetero_____ accretion theory. This theory holds that during the accretion process the solar nebula _____cooled_____. This process caused particles of _____metals_____ and _____metal oxides_____ to cool first, and the protoplanets began by forming _____metallic_____ cores. As the solar system continued to _____cool_____, the _____silicate_____ condensed. The protoplanets then added _____silicates_____ mantles and _____crust_____ to the already-formed _____metallic_____ cores. The heat generated by this in-falling matter also caused the process of _____differentiation_____, the separation of low-density and high-density materials, to continue. According to this theory, the atmosphere of Earth formed _____later_____ and was not produced entirely by _____outgassing_____. The theory suggests that _____icy_____ planetesimals, forming late in the solar nebula's life, crashed into Earth and formed the atmosphere and possibly the early oceans. These objects were probably formed in the _____outer_____ reaches of the solar system when encounters with the planet _____Jupiter_____ drove them into a collision course with Earth.

atmosphere	iron	radioactive
center	Jupiter	same
composition	later	silicate
cool	low	silicates
cooled	low-density	similar
crusts	melt	slowly
differentiation	metal oxides	small
electrostatic forces	metallic	space
evaporated	metals	speeds
gravity	nickel	sticky coatings
heat	outer	sank
heterogeneous	outgassed	sunk
homogeneous accretion	outgassing	surface
icy	planetesimals ✓	terrestrial
interior	protoplanets	

Self-Test

Select the one best answer for each question. Check your answers with the Answer Key and review when necessary.

1. Evidence of cratering on the moon and the other terrestrial planets suggests that Earth
   a. never experienced a similar kind of cratering.
   b. experienced a similar kind of cratering earlier in its history.
   c. will experience a similar kind of cratering in the future.
   d. will never experience a similar kind of cratering.

2. Because of Earth's mass, our planet's atmosphere
   a. contains mostly gases that are similar to those on the Jovian planets.
   b. contains the same percentages of gases that exist throughout the solar system.
   c. contains very little hydrogen and helium.
   d. is composed primarily of carbon dioxide.

3. The orbits of the planets
   a. all lie in approximately the same plane.
   b. lie perpendicular to the orbit of the Earth.
   c. are perfectly circular.
   d. are inclined to the ecliptic, according to the Titius-Bode rule.

4. Each of the planets in our solar system orbits the sun
   a. in circular orbits.
   b. in the direction opposite to which they rotate, clockwise as seen from the north (above).
   c. in the same direction.
   d. opposite in direction to the orbital motion of Earth.

5. The orbits of the planets lie mainly in the same plane because the planets
   a. formed from a large thin disk of material surrounding the new sun.
   b. were captured by an already-formed sun.
   c. were pulled out of the sun by the gravitational influences of a star passing nearby.
   d. formed at a variety of locations but were pulled into a single plane by the gravitational influence of the sun.

6. The solar system has relatively little gas and dust between the planets. The lack of this material is probably due to the

    a. "vacuuming up" of the material by the gravitational attraction of the planets.

    b. condensation properties of the Jovian planets.

    c. radiation pressure from the young sun.

    d. accretion processes of the terrestrial planets.

7. The solar nebula theory describes how the planets formed from

    a. the capture of planetesimals in orbit around a nearby star.

    b. the condensation of a cloud of pure hydrogen and helium.

    c. the accretion of high-density matter in the vicinity of the newly forming sun.

    d. a disk-shaped cloud of gas and dust around the protosun.

8. The terrestrial planets nearest the sun are

    a. large and composed of ice and hydrogen and helium gas.

    b. small, solid, and rocky.

    c. irregular in shape and composed mostly of ice and rock.

    d. spherical in shape with solid crusts and gaseous interiors.

9. The Jovian planets are made primarily of

    a. hydrogen and helium.

    b. methane and ammonia.

    c. liquid and liquid metallic hydrogen.

    d. icy rock in solid form.

10. Most asteroids revolve around the sun between the orbits of

    a. Jupiter and Saturn.

    b. Earth and Mars.

    c. Saturn and Uranus.

    d. Mars and Jupiter.

11. Objects made mostly of ice that form a large gaseous "tail" that is directed away from the sun at all times are called

    a. asteroids.

    b. meteors.

    c. meteorites.

    d. comets.

12. Formation of planetesimals by condensation refers to the process of

    a. meteorites colliding and sticking together.

    b. asteroids colliding and breaking apart.

    c. matter building one atom at a time.

    d. asteroid collisions and adhesion.

13. Particles sticking together to form larger particles describes the process of

    a. accretion.

    b. adhesion

    c. condensation.

    d. adherence.

14. The formation of planetesimals by the accretion of matter with the same general composition describes the process of

    a. heterogeneous accretion.

    b. homogeneous accretion.

    c. condensation.

    d. evaporative accretion.

15. The separation of low-density and high-density materials in a planet during the early stages of its formation is referred to as

    a. accretion.

    b. differentiation.

    c. homogeneous condensation.

    d. heterogeneous condensation.

Using What You've Learned

1. To understand the relative sizes of the planets, construct a scale model of the solar system. Start by establishing a scale for the diameters of the planets. Use Earth's diameter of 12,656 kilometers as equal to 1 inch. Then, using the data in the second column of the table "Planets: Physical Properties" at the bottom of page A18 in Appendix D of the textbook, calculate inch equivalents for all the other planets. To translate these figures into representations of the sizes of the planets, take a roll of adding machine paper tape and cut it into appropriate lengths to represent the diameters of the other planets. For example, Jupiter has a diameter of 142,400 kilometers, which is 11.18 times the diameter of Earth. Cut a piece of tape 11.2 inches long. Place this tape next to the tape representing Earth. Only by placing the strips of paper next to each other can you appreciate the relative sizes of the terrestrial and Jovian planets.

   Complete the scale representation of the solar system by cutting a strip of paper representing the sun. The sun has a diameter of about 1,400,000 kilometers, compared to a diameter of 12,656 kilometers for Earth. Divide the sun's diameter by Earth's diameter and you discover that the sun is approximately 111 times larger in diameter than Earth. Cut a piece of tape 111 inches long and place that next to the pieces representing the planets.

2. Repeat the above activity, this time comparing the distances of the planets from the sun. Use the information in Appendix D, "Planets: Orbital Properties," on page A19 of the textbook to determine the distance of each planet from the sun. Use the scale of 1 foot to 1 astronomical unit. According to this scale, Earth is 1 foot away from the sun. Although you can quickly see that Pluto would be 39 feet (39 astronomical units) away from the sun, you cannot appreciate how far that actually is until you actually unroll a 39-foot strip of adding machine tape. Cut a piece of tape for each of the distances given on page A19 and compare. (Be careful not to compare the lengths of tape in exercise 1 with the lengths of tape in this exercise. These exercises did not use the same scale.)

Answer Key

Matching

1.	e	12.	b
2.	m	13.	h
3.	f	14.	l
4.	a	15.	o
5.	v	16.	g
6.	p	17.	u
7.	s	18.	k
8.	j	19.	d
9.	i	20.	n
10.	r	21.	q
11.	c	22.	t

Completion

1. counterclockwise, counterclockwise, Venus, Uranus, planets, minor members, counterclockwise, same

   common plane, edge on, straight, horizontal, thin disk, orbits, thin disk, equators, slightly, Earth's

   two, Earth, small, solid, rocky, terrestrial, large, gaseous, solid, Jupiter, Jovian, hydrogen, helium

   moons, asteroids, comets, rocks, ice, gas, dust, comets, gas, dust, solar nebula theory

2. planetesimals, protoplanets, same, speeds, low, sticky coatings, electrostatic forces, planetesimals, gravity

   terrestrial, planetesimals, composition, homogeneous accretion, slowly, heat, space, similar, composition, radioactive, heat, melt, iron, nickel, sank, center, silicates, surface, low density, evaporated, outgassed, interior, atmosphere

   terrestrial, heterogeneous, cooled, metals, metal oxides, metallic, cool, silicates, silicate, crusts, metallic, differentiation, later, outgassing, icy, outer, Jupiter.

Self Test

(Page numbers refer to the textbook.)

1. b (Objective 1; page 357)

2. c (Objective 1; page 349)

3. a (Objective 2; page 347)

4. c (Objective 2; page 349)

5. a (Objective 2; pages 347–348; video program)

6. c (Objective 3; pages 362–363)

7. d (Objective 3; pages 347–348; video program)

8. b (Objective 4; pages 349–350)

9. c (Objective 4; pages 351–352)

10. d (Objective 5; page 355)

11. d (Objective 5; pages 355–356)

12. c (Objective 6; page 358)

13. a (Objective 6; page 359)

14. b (Objective 6; page 360)

15. b (Objective 6; pages 360–361)

LESSON
20

Planet Earth

Assignments

For the most effective study of this lesson, we suggest that you complete the assignments in the sequence listed below.

Before Viewing the Video Program

- Read the Overview and Learning Objectives for this lesson. Use the Learning Objectives to guide your reading, viewing, and thinking.

- Read Chapter 17, "The Earthlike Planets," section 17–1, pages 367–377, in the textbook.

- Read the Viewing Notes in this lesson.

View "Planet Earth" Video Program

After Viewing the Video Program

- Briefly note your answers to questions listed at the end of the Viewing Notes.

- Review all reading assignments for this lesson, especially the textbook Chapter 17 summary on page 403 and the Viewing Notes in this lesson.

- Write brief answers to review questions 1–3 at the end of textbook Chapter 17 to be certain you understand the text material.

- Complete the Review Activities in this guide to reinforce your understanding of important terms and concepts. Check your answers with the Answer Key at the end of this lesson and review when necessary.

- Take the Self-Test in this guide to measure your achievement of the Objectives. Check your answers with the Answer Key and review when necessary.

- Complete the Using What You've Learned Activities and any other activities and projects assigned by your instructor.

Overview

The previous lesson began the study of the solar system—the region of the universe that we would consider to be local. Our solar system consists of several types of objects: the sun, nine major planets, several dozen moons, and a multitude of smaller objects such as comets and asteroids. Planetary astronomers categorize the nine planets into two groups according to their similar surface features and internal characteristics. This lesson is the first of three that examine the terrestrial planets, beginning with Earth. (The Jovian planets are covered in Lessons 23 and 24.) The terrestrial planets have relatively high densities, hard surfaces, and few satellites. In order to understand the formation and evolution of these planets, we must compare their characteristics and history to Earth's. The more we know about Earth, the better we will be able to understand the forces that shaped the other terrestrial planets. This program is devoted to the planet we have studied for the longest time and in the greatest detail. The knowledge we gained and the processes we have employed in studying Earth will be used to study the other planets. Thus, this lesson provides the foundation for the study of the rest of the solar system and the application of what is called *comparative planetology*.

When Earth was formed by the accretion process, the heat generated by these impacts and the heat released as a result of the decay of radioactive elements caused the planet's interior to become molten. While Earth was in this molten state, heavier elements such as nickel and iron drained to the center, while the lighter silicates rose to the surface. These lighter compounds solidified and became Earth's crust. This process of differentiation, the separation of materials into layers depending on their density, is common to all the terrestrial planets. Earth, however, is unique among the planets in that it is the only planet that continues to have extensive activity in its solidified crust.

As this lesson explains, the heat in the middle layer, or mantle, of Earth keeps the material underneath the crust in constant motion, causing the crust to crack and form large sections. These sections, or plates, collide and slide past each other, producing earthquakes, volcanoes, mountain ranges, and deep ocean trenches. No other planet has a geology so driven. However, as we learn more about Earth's geologic activity, we also come closer to understanding why the other planets do not show the same plate tectonics.

Erosion and the motion of Earth's crust through plate tectonics have destroyed all signs of the planet's earliest geologic record. No crater remains that is more than a few tens of thousands of years old. Plate tectonics are also responsible for the development of Earth's atmosphere. Gases such as carbon dioxide and water vapor, released from the molten interior, altered the history of Earth. Water vapor, in an environment cool enough for it to condense and form liquid water, removed the carbon dioxide from the atmosphere and allowed Earth's surface to cool. Chemicals within that liquid water began to form complex molecules and duplicate themselves. Life had formed. Once this life was capable of photosynthesis, it produced extra oxygen, eventually allowing a layer of ozone to form in the upper atmosphere. This ozone layer protected the primitive life forms from harmful radiation and, in time, allowed life to move onto the land. Earth's atmosphere, from its earliest composition of gases left over from the solar nebula to its present composition of oxygen and nitrogen, is dynamic and critical in the support of life on this planet.

Today, however, the atmosphere that protected primitive life in the oceans and on land may be in danger because of the industrial processes in which that life is engaged. Carbon dioxide, once removed by liquid water, is now being returned to the atmosphere, possibly raising global temperatures. The ozone layer, screening harmful ultraviolet radiation, is being destroyed by chlorofluorocarbons (CFCs). The geological activity on Earth will continue regardless of the activity of the life that currently inhabits this planet. However, if intelligent life is to survive on this planet, we must understand the geological and atmospheric processes that affect it. As we study Earth to better understand the other terrestrial planets, we see that by studying the other terrestrial planets we can better understand processes that occur on Earth. In addition to gaining knowledge about the individual objects in our solar system, the combined knowledge helps us to understand the origin of the solar system itself.

Learning Objectives

When you have completed all assignments for this lesson, you should be able to:

1. Outline a four-stage history of Earth's formation. TEXTBOOK PAGE 368.

2. Describe a model of the Earth's interior and explain how we have come to develop that model. TEXTBOOK PAGES 368–369; VIDEO PROGRAM.

3. Explain the current theory about how the magnetic field of Earth is produced. TEXTBOOK PAGE 369; VIDEO PROGRAM.

4. Explain the theory of and evidence for plate tectonics and continental drift. TEXTBOOK PAGES 369 and 373–374; VIDEO PROGRAM.

5. Describe and explain the processes that have changed Earth's atmosphere from its earliest formation to the present day. TEXTBOOK PAGES 375–377; VIDEO PROGRAM.

6. Explain the cause and possible consequences of a depletion of Earth's ozone layer. TEXTBOOK PAGE 376; VIDEO PROGRAM.

7. Explain the concept of global warming and describe the possible consequences. TEXTBOOK PAGES 375–376; VIDEO PROGRAM.

8. Describe the effects life on Earth may have on Earth's environment. TEXTBOOK PAGE 376; VIDEO PROGRAM.

Viewing Notes

In this video program, planetary scientists, geophysicists, and geologists outline the formation and evolution of Earth. They also discuss how our understanding of the process that formed our planet contributes to our understanding of the other planets of our solar system and how, as we learn more about those other planets, our knowledge of Earth grows.

The first segment describes the early evolution of Earth. Computer animation illustrates the formation of the planet by accretion and its eventual differentiation into a sphere with an iron core surrounded by a mantle and crust. Planetary scientists describe the dynamo effect, the process by which the motion of the molten iron in the core generates Earth's magnetic field and how earthquakes and data gathered from seismometers have helped planetary scientists learn more about Earth's interior. As explained in the video program, earthquakes occur when plates "floating" on the mantle move. The motion is a result of convection, which causes the mantle material to flow continuously.

The video program then considers plate tectonics, one of the dynamic phenomena still shaping Earth today. Computer graphics illustrate how convection broke up the supercontinent of Pangaea and ultimately formed the continents as we know them today. This segment also explains the dynamic phenomena of plate tectonics, seafloor spreading, midocean ridges, subduction zones, and volcanism and their ongoing role in shaping Earth. Geophysicists and a planetary scientist briefly compare Earth to the other terrestrial planets, explaining the reason for the relatively small number of visible craters on Earth and noting the apparent absence of plate tectonics on other planets.

The video program also covers Earth's most unique features—its liquid water and atmosphere—and describes how water was formed and how our atmosphere evolved. This last segment of the program explains the changes that occurred in Earth's oceans and atmosphere that enabled life to develop on the planet. An atmospheric scientist and a geological oceanographer describe the role of carbon dioxide in Earth's atmosphere and oceans and how the oceans, over millions of years, dissolved much of the carbon dioxide, forming carbonates and carbonate rock (limestone).

This segment also describes the greenhouse effect and the ozone layer and their role in allowing life to develop on Earth, and it explains how the oceans and atmosphere evolved to contain sufficient oxygen to allow organisms to live on land.

As you watch the video program, consider the following questions:

1. How did Earth form and what processes shaped it to its present form?

2. How are earthquakes and seismometers used to study the structure of Earth's interior?

3. How do plate tectonics reflect convection in the mantle and how are plate tectonics related to continental drift and seafloor spreading?

4. Why is more cratering visible on the other terrestrial planes and on the moon than on Earth?

5. What characteristics make Earth's surface unique among the terrestrial planets? Why?

6. How did Earth's atmosphere develop?

7. What changes occurred in Earth's atmosphere and oceans that enabled organisms to live on land?

8. What is ozone and why is it necessary in order for life to exist on the surface of the Earth?

Review Activities

Matching

Match the terms listed below with the definitions that follow. Check your answers with the Answer Key and review any terms you missed.

_I_ 1. comparative planetology

_C_ 2. mantle

_b_ 3. plastic

_L_ 4. dynamo effect

_H_ 5. basalt

_n_ 6. midocean rise

_J_ 7. midocean rift

_k_ 8. rift valley

_e_ 9. plate tectonics

_d_ 10. greenhouse effect

_G_ 11. primeval atmosphere

_a_ 12 Pangaea

_m_ 13. seafloor spreading

_f_ 14. continental drift

a. name of the single land mass from which the current continents were formed

b. a material with the properties of a solid but capable of flowing under pressure

c. layer of dense rock and metal oxides between the molten core and Earth's surface

d. process by which an atmosphere rich in carbon dioxide entraps heat and raises the temperature of a planetary surface

e. constant destruction and renewal of Earth's surface resulting from movement of sections of crust

f. observed motion of the large land masses sitting on top of the crustal plates as a result of plate tectonics

g. Earth's first air

h. dark igneous rock characteristic of solidified lava

i. study of planets in relation to one another

j. chasm that splits the midocean rises where crustal plates move apart

k. long, straight, deep valley produced by the separation of crustal plates

l. process by which Earth's magnetic field is generated in the conducting material of its molten core

m. process by which molten material from the mantle emerges from the midocean rifts, pushing the continents further apart

n. undersea mountain range that pushes up from the seafloor in the center of an ocean

Completion

Fill each blank with the most appropriate term from the list for that paragraph. A term may be used once, more than once, or not at all. Check your answers with the Answer Key and review when necessary. If a question requires two or more answers in succession, they may be in any order, unless the question indicates otherwise.

1. The crust of Earth is divided into numerous plates. Movement of these plates, which results from _convection_ in Earth's _mantle_, causes the _continents_ to separate. Evidence of this

separation is indicated by the presence of mountain ranges, or _midocean rises_, in the middle of the oceans. In the center of these mountain ranges are _rifts_, from which _molten_ rock rises up, causing the seafloor to _spread_ outward. The strongest evidence supporting the phenomenon of _seafloor_ is found in the residual _magnetism_ in the solidified rock on the seafloor. Some of this material is found to have a _magnetic_ polarity _reversed_ from Earth's present magnetic field. These regions, on both sides of the _rifts_, show _symmetric_, parallel, and _alternating_ bands of magnetism that demonstrate that Earth is creating new _crust_, causing the _seafloor_ to spread and push the continents apart.

alternating	midocean rises
continents	molten
convection	reversed
crust	seafloor
magnetic	seafloor spreading
magnetism	spread
mantle	symmetric
midocean rifts	

2. Earth's first atmosphere, sometimes called the _primeval_ atmosphere, was thought to consist of gases originally belonging to the _nebula_. These original gases included _hydrogen_, _helium_, _ammonia_, and _methane_. More recent studies of the origin of Earth, however, indicate that temperatures on the newly forming Earth were hot enough to release gases such as _carbon dioxide_ and _water vapor_, thus permitting the first atmosphere to become rich in these gases. As liquid water collected on Earth's surface, the water began to dissolve the _carbon dioxide_ thus reducing the _greenhouse effect_ and lowering Earth's surface _temperature_. Earth's atmosphere continued to change. The primeval gases of _ammonia_ and _methane_ were destroyed by _ultraviolet_ radiation. With the reduction of the gas _greenhouse_ being dissolved into the oceans, the atmosphere of Earth became rich in the gas _nitrogen_. Another important gas that allowed living organisms to venture out of the oceans and onto the land eventually appeared in Earth's atmosphere. This gas, known as

1. % of light an object reflects

2. How squashed Earth is
   equator vs Poles Pole.

Ozone , prevented ultraviolet radiation from reaching Earth's surface.

Today, we face two possible problems because of human activity on Earth. First, we are concerned that the amount of carbon dioxide being released into the atmosphere is increasing. Since the amount of _infrared_ radiation leaving the surface of Earth would be reduced, the temperature of Earth's surface would _increase_ . If this pattern of _global warming_ were to continue unabated, the surface temperature would be such that the ice caps would begin to _melt_ and ocean levels would begin to _rise_ . Second, we are concerned about the destruction of the _ozone layer_ in the Earth's atmosphere. Modern chemicals called _CFC_ destroy the _ozone layer_ when they leak into the atmosphere. If this layer were significantly reduced, the amount of _ultraviolet_ reaching the surface of Earth would _increase_ and have a detrimental effect on life on this planet.

ammonia	methane
carbon dioxide	nitrogen
chlorofluorocarbons	ozone
decrease	ozone layer
freeze	primeval
global warming	rise
greenhouse effect	solar nebula
helium	temperature
hydrogen	ultraviolet
increase	ultraviolet radiation
infrared	water vapor
melt	

Self-Test

Select the one best answer for each question. Check your answers with the Answer Key and review when necessary.

1. Iron and nickel drained to the center of Earth, while at the same time, the lighter silicates rose to the surface, solidified, and formed a thin crust. This process in known as
   a. differentiation.
   b. sedimentation.
   c. liquefaction.
   d. distillation.

2. Which of the following sequences best describes the four-stage history of Earth's formation and evolution?
   a. differentiation, surface evolution, cratering, basin flooding
   b. condensation, cratering, basin flooding, differentiation
   c. differentiation, cratering, basin flooding, surface evolution
   d. cratering, condensation, accretion, surface evolution

3. Earth's core has a solid center and liquid outer parts. Evidence for the liquid component of the core comes from
   a. gravity studies by spacecraft.
   b. seismic wave studies.
   c. physical measurement by subterranean probes.
   d. oil well drillings.

4. Between the crust and liquid core lies a solid material that behaves as a plastic, flowing under high pressure. This material makes up Earth's
   a. mantle.
   b. crust.
   c. inner core.
   d. continental shelf.

5. The differential rotation between Earth's liquid core and the planet itself produces Earth's

   a. aurora.

   b. seismic activity.

   c. plate tectonics.

   d. magnetic field.

6. Earth's magnetic field is thought to result from the dynamo effect, which is

   a. motion in the liquid core that generates seismic waves.

   b. motion in the liquid core that generates electric currents.

   c. thermal cooling in the liquid core that generates temperature gradients.

   d. seismic convection in the liquid core that generates liquefaction.

7. Earth's crust is divided into large sections that slowly move about as the result of

   a. seismic activity in the crust.

   b. convection currents in the core.

   c. convection currents in the mantle.

   d. seafloor separation.

8. One of the best ways to locate the crustal plate boundaries is to identify

   a. large mountain ranges on the continents.

   b. lakes, major rivers, and oceans.

   c. the outline of the continental shelf.

   d. the locations of earthquakes.

9. Earth's primeval atmosphere probably consisted of all of the following gases **except**

   a. oxygen.

   b. carbon dioxide.

   c. water vapor.

   d. nitrogen.

10. The carbon dioxide in Earth's early atmosphere was produced when large amounts of gases were released during the accretion process when Earth's surface temperature was very hot. Most of the carbon dioxide remained in Earth's atmosphere until it was removed when

    a. ultraviolet radiation caused the gas to disassociate into carbon and oxygen.

    b. volcanic activity produced chemicals that neutralized the gas.

    c. liquid water formed on Earth's surface and dissolved the carbon dioxide causing it to react with other compounds to form limestone and other sediments.

    d. infrared radiation heated Earth's atmosphere and evaporated the carbon dioxide.

11. The form of electromagnetic radiation that the ozone layer prevents from reaching Earth's surface is

    a. X rays.

    b. visible light.

    c. ultraviolet radiation.

    d. infrared rays.

12. Ozone is actually a compound of the element

    a. hydrogen.

    b. oxygen.

    c. nitrogen.

    d. helium.

13. The temperature of Earth is much warmer and more uniform than it would be if Earth did not have an atmosphere. The gas mostly responsible for trapping heat in Earth's atmosphere is

    a. carbon dioxide.

    b. oxygen.

    c. helium.

    d. nitrogen.

14. The greenhouse effect best describes the process by which
    a. ultraviolet radiation is filtered from sunlight.
    b. infrared radiation is reflected off the top of Earth's atmosphere.
    c. carbon dioxide is dissolved by liquid water.
    d. heat is trapped by the atmosphere, thus increasing the temperature at the surface of Earth.

15. Free oxygen in Earth's atmosphere was produced by
    a. chemical reactions taking place in the oceans.
    b. living organisms during photosynthesis.
    c. chemical reactions taking place on land.
    d. the decay of early plant life.

16. A major concern of scientists today is the overproduction of chemicals known as CFCs. These chemicals
    a. pollute ocean waters and destroy plant life.
    b. destroy the ozone layer in Earth's atmosphere.
    c. enrich the soil, producing an overabundance of plant life and causing the loss of soil nutrients.
    d. do nothing harmful. Scientists are misguided in their conclusions.

Using What You've Learned

1. Diagram the internal structure of Earth, identifying the four layers. Continue the diagram to include the five major regions of Earth's atmosphere. List the altitude of each region and identify where interplanetary space begins.
    a. In which layer of the atmosphere does weather occur?
    b. What is the hydrosphere?
    c. What is the lithosphere?

2. Answer the following questions:

   a. Why is the sky blue?

   b. Why does the sun sometimes appear larger at the horizon, just before sundown, than when it is overhead?

   c. How do halos form around the sun and the moon?

   d. What are mirages and how do they occur?

3. Visit the geology and earth science exhibits at a museum to learn more about Earth.

4. You discover a new satellite of Earth. You send radar waves to it and receive the radar exactly 2 seconds after the signal is transmitted. What is the distance to the satellite in kilometers?

5. Make a list of the scientific evidence that supports the existence of the supercontinent of Pangaea. For example: Identify landforms and rock deposits that are the same, but lie on two different continents. Identify animal and plant species (can be fossils) that exist on two continents separated by an ocean.

Answer Key

Matching

1.	i	8.	k
2.	c	9.	e
3.	b	10.	d
4.	l	11.	g
5.	h	12.	a
6.	n	13.	m
7.	j	14.	f

Completion

1. convection, mantle, continents, midocean rises, midocean rifts, molten, spread, seafloor spreading, magnetism, magnetic, reversed, midocean rifts, symmetric, alternating, crust, seafloor

2. primeval, solar nebula, hydrogen, helium, methane, ammonia, carbon dioxide, water vapor, carbon dioxide, greenhouse effect, temperature, ammonia, methane, ultraviolet, carbon dioxide, nitrogen, ozone

   infrared, increase, global warming, melt, rise, ozone layer, chlorofluorocarbons, ozone layer, ultraviolet radiation, increase

Self-Test

(Page numbers refer to the textbook.)

1. a (Objective 1; page 368; video program)

2. c (Objective 1; page 368)

3. b (Objective 2; page 369; video program)

4. a (Objective 2; page 369; video program)

5. d (Objective 3; page 369; video program)

6. b (Objective 3; page 369; video program)

7. c (Objective 4; page 369; video program)

8. d (Objective 4; pages 374–375)

9. a (Objective 5; page 375)

10. c (Objective 5; pages 375–376; video program)

11. c (Objective 6; page 376)

12. b (Objective 6; page 376)

13. a (Objective 7; pages 375–376; video program)

14. d (Objective 7; page 375; video program)

15. b (Objective 8; page 376; video program)

16. b (Objective 8; page 376)

L E S S O N
21

The Moon and Mercury

Assignments

For the most effective study of this lesson, we suggest that you complete the assignments in the sequence listed below.

Before Viewing the Video Program

- Read the Overview and Learning Objectives for this lesson. Use the Learning Objectives to guide your reading, viewing, and thinking.

- Read Chapter 17, "The Earthlike Planets," sections 17-2 and 17-3, pages 377–386, in the textbook.

- Read Background Notes 21A, "Radar Astronomy," and 21B, "Synchronous Rotation," following Chapter 26 in this guide.

- Read the Viewing Notes in this lesson.

View "The Moon and Mercury" Video Program

After Viewing the Video Program

- Briefly note your answers to the questions listed at the end of the Viewing Notes.

- Review all reading assignments for this lesson, especially the textbook Chapter 17 summary on page 403, the Background Notes 21A and 21B, and the Viewing Notes in this lesson.

- Write brief answers to review questions 4–6 at the end of textbook Chapter 17 to be certain you understand the text material.

- Complete the Review Activities in this guide to reinforce your understanding of important terms and concepts. Check your answers with the Answer Key at the end of this lesson and review when necessary.

- Take the Self-Test in this guide to measure your achievement of the Objectives. Check your answers with the Answer Key and review when necessary.

- Complete the Using What You've Learned activities and any other activities and projects assigned by your instructor.

Overview

In our exploration of the solar system, we now move to a close-up study of the object nearest to Earth, the moon. Of all the objects in the night sky, it is perhaps the most familiar. Throughout human history, its changing appearance has stimulated numerous mythologies. Even today, poetry and songs celebrate the moon as a romantic object, which may seem surprising in that our closest celestial neighbor is a cold and desolate world.

Galileo, using his small telescope, was the first to identify features on the lunar surface. He correctly labeled mountains and craters. He thought the smooth, dark areas were oceans of water, which he called *mare*. Later, larger telescopes revealed these "seas" to be plains of rock. Without an atmosphere and without evidence of running water on the surface, the moon is a lifeless, unchanging world.

The Apollo manned lunar missions confirmed that the moon and Earth were very different. Samples brought back by these missions showed that moon rocks lacked water and other low-temperature volatiles. Seismometers left on the surface of the moon revealed that the interior of the moon is quite cold and does not have a liquid core. Without a warm interior, the crust of the moon is locked in place, and plate tectonics do not occur.

For planetary scientists studying the moon, one of the major questions is, "Why are there such major differences between Earth and the moon?" The search for the answer to this question is like a detective story. Using observations, results of experiments, analysis of lunar samples, and computer modeling and simulations, planetary scientists have developed a model for the formation and evolution of the moon. The first part of this lesson examines that model and compares it with Earth's history. As you will learn, the moon is a world that followed an evolutionary path completely different from Earth's.

Although Earth and the moon are very different, another object in our solar system appears, at least at first glance, to be very similar to the moon. Mercury is a heavily cratered planet. The number and distribution of its craters are similar to the moon's. Mercury also has large lava-filled plains like the lunar maria. In fact, if one did not know the difference, photographs of the moon and Mercury could easily be confused. However, the gravitational influence Mercury had on the Mariner 10 spacecraft as it flew by the planet revealed that its interior was distinctly different from that of the moon.

In order for Mercury to influence the spacecraft in the way it did, Mercury had to be extremely dense; in fact, the overall density of Mercury is approximately equal to that of Earth. For Mercury to be so dense, it must have an iron core at least two-thirds the diameter of the planet. Furthermore, some of that core is liquid. This conclusion was drawn after Mariner 10 detected a weak magnetic field about the planet. It seems that Mercury is like the moon on the outside, but more like Earth on the inside.

Our study of the moon, Mercury, and the other planets helps us to understand the processes involved in the formation and evolution of Earth. By comparing similarities and differences of these other worlds, we can see how unique and special Earth truly is.

Learning Objectives

When you have completed all assignments for this lesson, you should be able to:

1. Identify the major features and characteristics of the surface of the moon and describe the various processes that formed them. TEXTBOOK PAGES 377–379; VIDEO PROGRAM.

2. List and describe the various geological findings (surface and interior) made by spacecraft missions (manned and unmanned) to the moon. TEXTBOOK PAGES 377–382; VIDEO PROGRAM.

3. Describe the various theories about the origin of the moon and state the advantages and disadvantages of each. TEXTBOOK PAGES 382–383; VIDEO PROGRAM.

4. Describe the surface features of Mercury as seen by the Mariner 10 spacecraft and explain their origin. textbook pages 383–386; VIDEO PROGRAM.

5. Describe a model for the interior of Mercury. TEXTBOOK PAGE 385; VIDEO PROGRAM.

6. Describe the radar observations of Mercury and cite advantages of radar observations over traditional Earth-based optical observations. TEXTBOOK PAGE 385; BACKGROUND NOTES 21A and 21B; VIDEO PROGRAM.

Viewing Notes

Two worlds, similar in outward appearance, but quite different inside, are the subject of this video program, "The Moon and Mercury." The video program begins by looking at the moon, the celestial object closest to Earth and one that has long been observed by humans. Edwin Krupp, an archaeoastronomer, describes the importance of the moon to ancient cultures. Other astronomers explain how Galileo's early telescopic observations of the moon changed western civilization's perception of the moon. Next, the video program illustrates what we now know about the basic features of the moon—the heavily cratered highlands and the lava-filled lowlands, the maria—and describes how these features were formed. This segment also explains synchronous rotation and why the same side of the moon always faces Earth. It includes photographs taken by Soviet spacecraft in 1959 of the back of the moon and an explanation of why the two hemispheres have a different appearance.

The moon is unique in that it is the only extraterrestrial place that humans have personally explored and from which we have actual samples. NASA footage of the first Apollo landing on the moon in 1969 and of several lunar walks and explorations is accompanied by planetary astronomers' descriptions of what we have learned about the moon's structure and composition from seismometers placed on the moon and from lunar samples brought back by astronauts.

The video program's examination of the moon ends with descriptions of different hypotheses (fission, condensation, and capture) of how the moon was formed and discussion of the failure of most of them to account for specific facts about the moon. The large-impact hypothesis of moon formation is the one that is currently most accepted, and the co-developer of that theory, William K. Hartman, a senior scientist at the Planetary Institute, explains the theory. Computer simulation illustrates the large-impact hypothesis and how the moon might have been created when a Mars-sized object crashed into Earth, sending mantle material into space, where it ultimately condensed into the moon.

From the moon, the celestial object closest to Earth, the video program moves to Mercury, the planet closest to the sun. Unlike the moon, Mercury is difficult to observe through optical telescopes because it is small and so close to the sun. Planetary astronomers describe our early limited knowledge of Mercury. Most of what we know about Mercury has been learned fairly recently, from radar studies and from the Mariner 10 spacecraft, which photographed Mercury in 1974–75. Those photographs, shown in this segment, reveal a cratered surface remarkably similar to the moon's, along with maria, also similar to those on the moon but lighter in color, and scarps (tall cliffs), which are not present on the moon. Planetary astronomers describe those features and their formation and compare Mercury's structure to the moon's. Mariner 10 also detected a magnetic field and large iron core; and astronomers explain how, on the inside, Mercury is similar to Earth. Radar studies, which have mapped the entire surface of Mercury, have provided much of our current knowledge of the planet's surface, its composition, and its possible patches of water ice at the poles. Radar images of Mercury illustrate the value of this tool for studying the inner solar system.

As you watch the video program, consider the following questions:

1. What features on the lunar surface were first identified by telescopes, and what actions produced those features?

2. Why has the moon not been reshaped by erosion and plate tectonics?

3. What evidence suggests that the moon is asymmetric? How does this asymmetry cause the near and far sides of the moon to appear so different?

4. How does the large-impact hypothesis of the formation of the moon account for its differences in chemical composition and density when compared with Earth?

5. Why is optical observation of Mercury so difficult?

6. What did we know about Mercury before any spacecraft mission?

7. What are the similarities and differences between the surfaces of Mercury and the moon?

8. What information have radar studies of Mercury provided that could not be obtained by Earth-based optical telescopes?

Review Activities

Matching

Match the terms listed below with the definitions that follow. Check your answers with the Answer Key and review any terms you missed.

__D__ 1. ejecta

__i__ 2. rays

__e__ 3. vesicular basalt

__b__ 4. mare

__K__ 5. anorthosite

__H__ 6. breccia

__n__ 7. albedo

__g__ 8. early heavy bombardment

__f__ 9. mascon

__c__ 10. fission hypothesis

__j__ 11. condensation hypothesis

__L__ 12. capture hypothesis

__a__ 13. large-impact hypothesis

__m__ 14. lobate scarp

__p__ 15. radar astronomy

__o__ 16. synchronous rotation

a. theory that the moon formed from debris ejected during a collision between Earth and a large planetesimal

b. a lunar lowland filled by successive flows of dark lava

c. theory that the moon and Earth formed when a rapidly rotating protoplanet split into two pieces

d. pulverized rock scattered by meteorite impacts on a planet

e. rock with trapped bubbles formed by solidified lava

f. concentrations of mass found immediately below the maria on the moon

g. intense cratering that occurred during the first 500 million years in the history of the solar system

h. rock composed of fragments of earlier rocks bonded together

i. white streamers radiating from some lunar craters

j. theory that Earth and the moon condensed simultaneously from the same cloud of material

k. rock of aluminum and calcium silicates found in the lunar highlands

l. theory that Earth's moon formed elsewhere in the solar nebula and was later captured by Earth

m. a curved cliff such as those found on Mercury

n. the ratio of the light reflected from an object divided by the light that hits the object; this ratio equals 0 for perfectly black and 1 for perfectly white

o. identical rotation and revolution periods

p. use of artificially produced radio signals reflected off objects and analyzed according to the time delays and Doppler shifts of the return signal

Completion

Fill each blank with the most appropriate term from the list for that paragraph. A term may be used once, more than once, or not at all. Check your answers with the Answer Key and review when necessary. If a question requires two or more answers in succession, they may be in any order, unless the question indicates otherwise.

1. Upon viewing the moon through a telescope, one can identify the moon's two types of terrain. The light areas, or _Highland_, are heavily _cratered_. The other type of terrain on the moon can be described as large, relatively _Smooth_ plains of _darker_ material. Galileo believed these plains were _seas_ and therefore named them _maria_. The relative ages of these different terrains can be determined by counting the number of _craters_ in each. Such a count indicates that the _lowlands_ are younger than the _Highlands_.

   The exact composition of these two areas was determined when lunar samples were returned by the manned _Apollo_ missions. Samples from the maria were shown to contain _vesicle basalt_, which is a hardened form of _lava_. This finding confirms the theory that the maria are the result of _Impacts_ that left enormous craters that eventually filled with molten rock. The other lunar area, the _Highlands_, consists of rocks known as _Anorsite_. These rocks are composed of silicates of _Aluminum_ and _Calsium_. Their color is _lighter_, and their density is _lower_ than that of the _basalt_ found in the lowlands of the moon.

aluminum

anorthosite

Apollo

basalt

calcium

cratered

craters

darker

highlands

impacts

lava

lighter

lower

lowlands

maria

seas

smooth

vesicular basalt

2. Astronomers have considered several different hypotheses to explain the origin of the moon. One hypothesis was that Earth, in its early history, rotated much more _rapidly_ than it does today. Because of this rotation, the moon _broke away_ from Earth. To explain why the moon has so little _iron_, this hypothesis said that this major event occurred after Earth had _differentiated_. One of the problems of this _fission_ hypothesis was that it could not explain the differences in _composition_ between Earth and moon rocks. Furthermore, the Earth-moon system should contain a great deal more _angular momentum_ than is currently observed.

A second hypothesis was that the moon and Earth formed from the same cloud of material in the _solar nebula_. The moon and Earth _condensed_ simultaneously as twin planets. This _condensation_ hypothesis did not remain plausible because of the differences in the _chemical_ compositions and _density_ of the moon and Earth. Evidence of this difference is found in the fact that the moon is very poor in _volatiles_. In other words, the moon has little material that is easily _evaporated_, such as _water_ bound up in its rocks, while Earth is abundant in rocks that contain _water_.

A third hypothesis for the origin of the moon was that the moon formed _elsewhere_ in the solar system and was _captured_ by Earth. This hypothesis, the _capture_ hypothesis, solves a few problems. It allows for the moon to have significant differences in _composition_ and _density_. It also permits Earth to rotate as _slowly_ as it does today. Even so, this hypothesis has its problems. There must be a very special set of circumstances for the Earth to _capture_ the moon. The speed and _trajectory_ of the moon had to be such that when it approached Earth the moon would enter into _orbit_ around Earth and not be shattered by _tidal_ forces. Since the moon is so large, and with the numerous _trajectories_ that are possible, it is highly unlikely that the specific set of circumstances allowing a successful _capture_ would have been met.

At present, the most widely accepted hypothesis for the origin of the moon in called the _large impact_ hypothesis. This hypothesis suggests that, during Earth's early history, an object, about as large as _Mars_, _collided_ with Earth. Material from Earth's _mantle_ was blown into space and eventually condensed to form the _moon_. This explanation could account for the present amount of _angular momentum_ in the Earth-moon system. Furthermore, if Earth had already _differentiated_, it would explain why the moon has so little _iron_. The heat generated may very well have been enough for _water_ and other _volatiles_ to have been removed from the material that formed the moon.

angular momentum	differentiated	orbit
broke away	differentiation	rapidly
capture	elsewhere	slowly
captured	evaporated	solar nebula
chemical	fission	tidal
collided	iron	trajectories
composition	large-impact	trajectory
condensation	mantle	volatiles
condensed	Mars	water
density	moon	

3. The surfaces of Mercury and the moon resemble each other. They appear to have a similar number of _craters_. Both objects also have _maria_. On Mercury, however, the _maria_ are not as noticeable because they are filled with a _lighter_ -colored _lava_ as opposed to a _darker_ -colored _lava_ on the moon. Although many features on Mercury are similar to those on the moon, there is one noticeable difference. Mercury has large curved cliffs called _lobate scarps_, which do not have any parallels on the lunar surface. These cliffs are believed to be the result of _shrinkage_ in the crust of Mercury as the _interior_ of the planet _cooled_. Although not visible in the _Mariner 10_ photographs, radar observations of Mercury indicate the planet may have water ice at its _poles_. Signals sent from Earth in 1991 reflected off Mercury and returned as a _bright_ image. Analysis of this image indicated the

presence of ___*water ice*___. It is believed that the ice is in the soil of Mercury
on the floors of deep ___*craters*___. These regions are forever hidden from the
intense heat of the sun's rays.

cooled	maria
craters	Mariner 10
darker	poles
interior	radar bright
lava	shrinkage
lighter	water ice
lobate scarps	

Self-Test

Select the one best answer for each question. Check your answers with the Answer
Key and review when necessary.

1. Large, dark areas that cover a good portion of the side of the moon facing
   Earth are referred to as lunar

   a. maria.

   b. craters.

   c. escarpments.

   d. mascons.

2. The craters on the lunar surface were produced by

   a. volcanic activity.

   b. subsurface erosion due to running water.

   c. impacts by asteroids and meteorites.

   d. excavation by aliens.

3. The paths of orbiting spacecraft were slightly perturbed due to gravitation irregularities over the lunar seas. These irregularities are due to

   a. poor spacecraft maintenance.

   b. mascons located beneath the surface of the lunar seas.

   c. extremely dense and active volcanoes on the lunar surface.

   d. the uneven cratering between the lunar highlands and lunar maria.

4. Lunar rocks found in the highlands are

   a. older than rocks in the maria and composed mostly of basalt.

   b. younger than rocks in the maria and composed of aluminum and calcium silicates.

   c. older than rocks in the maria and composed mostly of aluminum and calcium silicates.

   d. younger than rocks in the maria and composed mostly of basalt.

5. The large-impact hypothesis proposes that the moon formed as a result of

   a. a large amount of material breaking off Earth due to Earth's rapid rotation.

   b. two large asteroids colliding, shattering, and condensing to form the moon.

   c. a young moon colliding with Earth, forming a ring system that later condensed to form the moon we see today.

   d. a Mars-sized object colliding with Earth and ejecting material from Earth's mantle, some of which fell back to Earth and the rest condensed to form the moon.

6. A problem with the fission hypothesis for the formation of the moon is the fact that

   a. the Earth-moon system has less angular momentum than would be expected from such a formation scenario.

   b. Earth and the moon have similar densities.

   c. it requires too many coincidental events.

   d. the moon is rich in volatiles.

7. The craters on Mercury are the result of

a. impacts by asteroids and meteorites.

b. volcanic activity.

c. subsurface erosion.

d. solar wind erosion.

8. Lobate scarps on Mercury were probably caused by

a. meteorite impacts.

b. shrinkage of the surface as the interior of the planet cooled.

c. surface erosion by flooding.

d. wind erosion.

9. Evidence for Mercury's iron core comes from

a. large impact craters.

b. unusually low overall density for a planet of its size.

c. unusually high overall density for a planet of its size.

d. large lava-filled maria.

10. The weak (although stronger than expected) magnetic field around Mercury detected by the Mariner 10 spacecraft indicates

a. a solid core.

b. solar flare activity.

c. the interior of the planet is similar to that of the moon.

d. the interior of the planet is similar to that of Earth.

11. Radar observations of Mercury were first used to determine the planet's

a. rotation rate.

b. chemical composition.

c. period of revolution about the sun.

d. inclination of its orbit.

12. Radar bright patches were observed at the north and south poles of Mercury. These patches suggest the presence of

    a. carbon dioxide gas.

    b. water ice.

    c. aluminum and calcium silicates in the soil.

    d. vesicular basalt in the planet's maria.

Using What You've Learned

1. Compare the similarities and differences between Mercury and the moon. Include not only the surface features but also their internal structures.

2. What problems would humans face if they tried to colonize Mercury?

3. For the next several lessons, you will be asked to locate the five naked-eye planets on the SC1 Equatorial Star Chart (provided for you by your instructor or available at your college bookstore), determine if they are visible at night, and then, using the constellation charts at the back of the *Horizons* text, go outside and try to find the planets. (You may also want to review "The Motion of Planets," pages 36–37, in the textbook.)

The SC1 Equatorial Star Chart

The Equatorial Star Chart shows the position of the stars and constellations situated within 60° north and south of the celestial equator. The celestial equator, if you recall, divides the sky into northern and southern hemispheres, just as the equator on Earth divides our planet into two hemispheres. In other words, if you stood on Earth's equator and looked up, the celestial equator (represented by the straight horizontal line through the center of the chart) would be directly overhead.

You will also notice the curved line running through the star chart. If you look closely, the line runs through the twelve constellations of the zodiac. This line is the ecliptic, the path the sun appears to follow through the sky over the course of a year. The small numbers along the ecliptic are celestial longitudes that you can use to locate the planets. Remember that the orbital planes of the planets lie close to the ecliptic plane and therefore the planets themselves are located on or near the ecliptic.

Locating the Stars

You have probably already identified several familiar constellations and located some of the brightest stars. A quick way of locating the stars is to use their coordinates.

The coordinates of stars are given in terms of their right ascension and declination (similar to longitude and latitude on a globe, respectively). The declination determines the position of the stars north (+) or south (-) of the celestial equator. If you look along the vertical line through the center of the chart, the line is marked off in degrees. It starts at 0° on the celestial equator and goes to 60° north (+) and 60° south (-) of declination. Right ascension is the horizontal coordinate, measured in hours, minutes, and seconds. It starts at the same vertical line at 0 hours. Running along the bottom of the chart, the right ascension goes from 0 to 12 hours on the left half of the chart, and from 12 to 24 hours on the right half of the chart.

Using the following table, "The Brightest Stars," locate several stars by using their coordinates of right ascension and declination. Draw an imaginary horizontal line through the declination coordinate on the star chart and an imaginary vertical line through the right ascension coordinate. The two lines should meet at the position of the star.

If you want to compare the chart with the appearance of stars in the night sky, hold your chart overhead, with the top of the chart toward the north, while you are facing south. Those stars that are near 0 hours right ascension will be nearly overhead in the early evening during November; those at 6 hours in February; those at 12 hours in May; and those at 18 hours in August.

BRIGHTEST STARS

Position (1980)

STAR	NAME	RIGHT ASCENSION	DECLINATION
1. α CMa A	Sirius	06 44.2	-16 42
2. α Car	Canopus	06 23.5	-52 41
3. α Boo	Arcturus	14 14.8	+19 17
4. α Cen A	Rigil Kentaurus	14 38.4	-60 46
5. α Lyr	Vega	18 36.2	+38 46
6. α Aur	Capella	05 15.2	+45 59
7. β Ori A	Rigel	05 13.6	-08 13
8. α CMi A	Procyon	07 38.2	+05 17
9. α Ori	Betelgeuse	05 54.0	+07 24
10. α Eri	Achernar	01 37.0	-57 20
11. β Cen AB	Hadar	14 02.4	-60 16
12. α Aql	Altair	19 49.8	+08 49
13. α Tau A	Aldebaran	04 34.8	+16 28
14. α Vir	Spica	13 24.1	-11 03
15. α Sco A	Antares	16 28.2	-26 23
16. α PsA	Fomalhaut	22 56.5	-29 44
17. β Gem	Pollux	07 44.1	+28 05
18. α Cyg	Deneb	20 40.7	+45 12
19. β Cru	Beta Crucis	12 46.6	-59 35
20. α Leo A	Regulus	10 07.3	+12 04

Locating the Planets

The following table provides the celestial longitudes of the sun and the five naked-eye planets at 10-day intervals. Select a day (near the present date) and locate the position of the sun on the ecliptic. (Remember that the celestial longitudes are the small numbers near the ecliptic on the chart.) Using the same dates, locate Mercury. Using the star charts in the back of the *Horizons* text, go outside (at sundown or just before sunrise) and try to observe the planet. If the planet is to the right of the sun's position on the ecliptic, it will be visible in the predawn sky, unless it is too close to the sun. When the planet is to the left of the sun, it will follow the sun in the sky and may be visible after sunset, if it is not too close to the sun. Of all the planets, Mercury is the most difficult to observe. Good luck hunting!

CELESTIAL LONGITUDES
OF THE SUN AND PLANETS

YEAR	DATE	SUN	MERCURY	VENUS	MARS	JUPITER	SATURN
1994	SEP 4	161	180	207	102	220	338
1994	SEP 14	171	195	216	108	222	337
1994	SEP 24	181	207	223	114	223	336
1994	OCT 4	191	215	227	120	225	336
1994	OCT 14	201	215	229	125	227	335
1994	OCT 24	211	204	226	130	229	335
1994	NOV 3	221	202	221	135	231	335
1994	NOV 13	231	213	216	140	234	335
1994	NOV 23	241	229	213	144	236	335
1994	DEC 3	251	245	215	147	238	336
1994	DEC 13	261	261	220	150	241	336
1994	DEC 23	271	276	227	152	243	337
1995	JAN 2	282	293	235	153	245	338
1995	JAN 12	292	309	245	151	247	339
1995	JAN 22	302	320	255	149	248	340
1995	FEB 1	312	319	266	147	250	340
1995	FEB 11	322	307	277	142	252	342
1995	FEB 21	332	307	289	138	253	343
1995	MAR 3	342	315	300	136	254	344
1995	MAR 13	352	328	312	134	255	345
1995	MAR 23	2	342	324	132	255	347
1995	APR 2	12	0	336	133	256	348
1995	APR 12	22	19	348	135	255	349
1995	APR 22	32	41	0	138	255	350
1995	MAY 2	41	60	12	141	254	351
1995	MAY 12	51	73	24	144	253	351
1995	MAY 22	61	78	36	149	252	352
1995	JUN 1	70	76	48	154	251	353
1995	JUN 11	80	70	60	158	249	353
1995	JUN 21	89	70	73	163	248	354
1995	JUL 1	99	77	85	169	247	354
1995	JUL 11	108	91	97	174	246	354
1995	JUL 21	118	110	110	180	245	354
1995	JUL 31	128	131	122	186	245	354
1995	AUG 10	137	151	134	192	245	354
1995	AUG 20	147	168	147	198	246	353
1995	AUG 30	156	182	159	205	247	352
1995	SEP 9	166	193	172	211	247	351
1995	SEP 19	176	200	184	218	249	350
1995	SEP 29	186	198	196	225	250	349

YEAR	DATE	SUN	MERCURY	VENUS	MARS	JUPITER	SATURN
1995	OCT 9	196	187	209	232	252	348
1995	OCT 19	205	187	221	239	253	348
1995	OCT 29	215	200	234	246	255	347
1995	NOV 8	226	216	246	253	257	347
1995	NOV 18	236	233	259	261	259	347
1995	NOV 28	246	249	271	268	261	347
1995	DEC 8	256	264	284	276	264	348
1995	DEC 18	266	280	296	284	266	348
1995	DEC 28	276	295	308	291	269	349
1996	JAN 7	286	305	320	299	271	349
1996	JAN 17	297	301	333	307	273	350
1996	JAN 27	307	290	345	315	275	351
1996	FEB 6	317	292	357	323	277	352
1996	FEB 16	327	301	9	331	279	353
1996	FEB 26	337	314	20	339	281	354
1996	MAR 7	347	329	32	346	283	356
1996	MAR 17	357	347	43	354	284	357
1996	MAR 27	7	6	53	2	286	358
1996	APR 6	17	27	63	10	287	0
1996	APR 16	27	45	72	18	287	1
1996	APR 26	36	57	80	25	288	2
1996	MAY 6	46	58	85	33	288	3
1996	MAY 16	56	53	88	40	288	4
1996	MAY 26	65	49	87	47	288	4
1996	JUN 5	75	51	82	55	287	5
1996	JUN 15	84	61	76	62	286	6
1996	JUN 25	94	76	71	69	285	6
1996	JUL 5	103	96	70	76	283	7
1996	JUL 15	113	118	73	83	282	7
1996	JUL 25	122	137	79	90	280	7
1996	AUG 4	132	154	87	96	279	7
1996	AUG 14	142	168	95	103	278	7
1996	AUG 24	151	179	105	110	278	6
1996	SEP 3	161	184	115	116	278	5
1996	SEP 13	171	179	126	123	278	5
1996	SEP 23	181	170	137	129	278	4
1996	OCT 3	190	172	149	135	279	3
1996	OCT 13	200	186	160	140	280	2
1996	OCT 23	210	203	172	146	281	1
1996	NOV 2	220	220	184	152	282	0

YEAR	DATE	SUN	MERCURY	VENUS	MARS	JUPITER	SATURN
1996	NOV 12	230	236	196	157	284	0
1996	NOV 22	240	252	209	162	286	0
1996	DEC 2	251	267	221	166	288	0
1996	DEC 12	261	281	233	171	290	0
1996	DEC 22	271	290	246	175	292	0
1997	JAN 1	281	283	259	179	295	1
1997	JAN 11	291	273	271	182	297	2
1997	JAN 21	302	277	284	185	300	2
1997	JAN 31	312	288	296	186	302	3
1997	FEB 10	322	301	309	186	305	4
1997	FEB 20	332	317	322	184	307	5
1997	MAR 2	342	334	334	182	309	6
1997	MAR 12	352	352	346	179	312	7
1997	MAR 22	2	13	359	174	313	9
1997	APR 1	12	30	11	171	315	10
1997	APR 11	21	40	24	168	317	11
1997	APR 21	31	38	36	167	319	13
1997	MAY 1	41	31	48	166	320	14
1997	MAY 11	51	29	61	167	321	15
1997	MAY 21	60	34	73	170	322	16
1997	MAY 31	70	46	86	173	322	17
1997	JUN 10	79	62	98	177	322	18
1997	JUN 20	89	82	110	181	323	19
1997	JUN 30	98	104	123	186	322	19
1997	JUL 10	108	124	135	191	321	20
1997	JUL 20	118	141	147	196	320	20
1997	JUL 30	127	154	159	202	319	20
1997	AUG 9	137	163	171	207	317	20
1997	AUG 19	146	166	183	213	316	20
1997	AUG 29	156	160	194	220	315	20
1997	SEP 8	166	153	206	226	313	19
1997	SEP 18	175	157	218	233	312	18
1997	SEP 28	185	173	229	240	312	17
1997	OCT 8	195	191	240	247	312	16
1997	OCT 18	205	208	251	254	312	15
1997	OCT 28	215	224	262	261	313	14
1997	NOV 7	225	240	272	269	313	14
1997	NOV 17	235	254	282	276	314	13
1997	NOV 27	245	267	290	284	316	13
1997	DEC 7	255	274	298	292	317	13

Answer Key

Matching

1. d
2. i
3. e
4. b
5. k
6. h
7. n
8. g

9. f
10. c
11. j
12. l
13. a
14. m
15. p
16. o

Completion

1. highlands, cratered, smooth, darker, seas, maria, craters, lowlands, highlands

   Apollo, vesicular basalt, lava, impacts, highlands, anorthosite, aluminum, calcium, lighter, lower, basalt

2. rapidly, broke away, iron, differentiated, fission, composition, angular momentum

   solar nebula, condensed, condensation, chemical, density, volatiles, evaporated, water, water

   elsewhere, captured, capture, density, composition, slowly, capture, trajectory, orbit, tidal, trajectories, capture

   large-impact, Mars, collided, mantle, moon, angular momentum, differentiated, iron, water, volatiles

3. craters, maria, maria, lighter, lava, darker, lava, lobate scarps, shrinkage, interior, cooled, Mariner 10, poles, radar bright, water ice, craters

Self-Test

(Page numbers refer to the textbook.)

1. a (Objective 1; page 379; video program)
2. c (Objective 1; pages 379–381; video program)
3. b (Objective 2; page 381)
4. c (Objective 2; page 379)
5. d (Objective 3; pages 382–383)
6. a (Objective 3; page 382)
7. a (Objective 4; page 385)
8. b (Objective 4; page 385)
9. c (Objective 5; video program)
10. d (Objective 5; video program)
11. a (Objective 6; Background Notes 21A and 21B)
12. b (Objective 6; page 385; video program)

LESSON
22

Venus and Mars

Assignments

For the most effective study of this lesson, we suggest that you complete the assignments in the sequence listed below.

Before Viewing the Video Program

- Read the Overview and Learning Objectives for this lesson. Use the Learning Objectives to guide your reading, viewing, and thinking.

- Read Chapter 17, "The Earthlike Planets, sections 17-4 and 17-5, pages 386–403, in the textbook.

- Review Background Notes 21A, "Radar Astronomy," following Lesson 26 in this guide.

- Read the Viewing Notes in this lesson.

View "Venus and Mars" Video Program

After Viewing the Video Program

- Briefly note your answers to the questions listed at the end of the Viewing Notes.

- Review all reading assignments for this lesson, especially the textbook Chapter 17 summary on pages 403–404 and the Viewing Notes in this lesson.

- Write brief answers to review questions 7–10 at the end of textbook Chapter 17 to be certain you understand the text material.

- Complete the Review Activities in this guide to reinforce your understanding of important terms and concepts. Check your answers with the Answer Key at the end of this lesson and review when necessary.

- Take the Self-Test in this guide to measure your achievement of the Objectives. Check your answers with the Answer Key and review when necessary.

- Complete the Using What You've Learned activities and any other activities and projects assigned by your instructor.

Overview

Venus and Mars have long fascinated both astronomers and casual observers. Venus, after the sun and moon, is the third brightest object in the night sky. Mars glows with a pale red color, distinct from all the other planets. From earliest times, these planets were included in lore and mythology throughout the world. The ancient Greeks and Romans linked Venus with beauty and fertility and Mars with war and power. With the invention of the telescope, observers of these two worlds could see the phases of Venus and the "canals" and seasons of Mars. Such observations increased astronomers' and the public's fascination with these planets.

Even seen through the largest telescopes, Venus appears as a bright, featureless disk against the dark sky. Early astronomers concluded that Venus must be covered with a layer of clouds and that these clouds were probably made of water vapor. It wasn't long before speculation led astronomers and others to believe that Venus was a tropical world with dense jungles and large dinosaurs roaming its surface. In contrast, Mars is the only planet whose surface can be seen in detail from Earth. Over one hundred years ago, astronomers observed that the Martian surface regularly changed in appearance; some concluded that those variations in color were the result of the growth of vegetation as Mars experienced its seasons. Scientists and nonscientists alike were intrigued by the thought that Mars and Venus might sustain life.

Although one of the main factors motivating planetary scientists to study Venus and Mars may have been the belief that life existed on these worlds, it is no longer a significant reason. The study of Venus and Mars is motivated by the need to understand the origin, evolution, and environments of these worlds in order to understand better the origin, evolution, and environment of Earth. It is comparative planetology, a theme that runs throughout these lessons on the solar system, that motivates scientists to study Venus and Mars.

Venus and Mars have been shaped by geologic and atmospheric forces similar to those on Earth. Venus and Mars are the two planets closest to Earth in the solar system. This fact enables planetary scientists to understand how only small differences in position in the solar system affect the evolution of a planet. For example, one would expect to see close similarities between Venus and Earth because of their similar size and density. The two planets, however, are quite different, and this lesson explains why they are different. In contrast, Mars is one-half the diameter of Earth and has an atmosphere only 1 percent as dense as Earth's, and yet its surface features are more similar to those on Earth than are the features on Venus. Furthermore, some of the Martian landforms are larger than anything found on any of the other terrestrial planets. Venus and Mars seem to be planets in the extreme. They have both more and less of what Earth has.

Learning Objectives

When you have completed all assignments for this lesson, you should be able to:

1. Describe the Pioneer, Magellan, and Venera spacecraft findings about the physical and geological features on the surface of Venus. TEXTBOOK PAGES 386–391; BACKGROUND NOTES 21A; VIDEO PROGRAM.

2. Describe the characteristics of the atmosphere of Venus and explain the role of the "greenhouse effect" in the environment on the surface of the planet. TEXTBOOK PAGES 391–394; VIDEO PROGRAM.

3. Compare and contrast Venus and Earth and explain why the two planets have such different environments. TEXTBOOK PAGES 386–394; VIDEO PROGRAM.

4. Describe and explain the origin of the geologic features found on the surface of Mars by spacecraft. TEXTBOOK PAGES 394–400; VIDEO PROGRAM.

5. Describe the characteristics of the Martian polar regions and compare them with Earth's. TEXTBOOK PAGES 399–402; VIDEO PROGRAM.

6. Describe the atmosphere of Mars and compare it with Earth's. TEXTBOOK PAGES 401–402; VIDEO PROGRAM.

7. Identify the two moons of Mars, describe their surface features, and explain what these features may reveal about the history of these two objects. TEXTBOOK PAGES 402–403.

Viewing Notes

This video program, through the use of computer-generated images from radar astronomy, provides close-up views of Earth's closest neighbors: Venus and Mars. A geophysicist and a planetary scientist introduce the two planets by noting some of the striking similarities between the two planets and Earth. The remainder of the video program then explores Venus and Mars, providing detailed information about each planet's geologic features and atmosphere.

The segments about Venus begin with optical telescope images of the cloud-shrouded planet while astronomers explain why Earth-based observation of Venus has been so difficult. Then, an astrophysicist describes how radar astronomy enabled planetary scientists to gather the first real data about Venus, including information on its rotation rate and direction. The video program next shows animation of how the Magellan spacecraft mapped Venus' surface with radar waves that penetrated the planet's cloud cover.

Data obtained from the Magellan program are converted by computer into three-dimensional aerial views of Venus. The video program shows computer-generated footage from NASA and the Jet Propulsion Laboratory of some of Venus' most dramatic features: Ishtar Terra, with Maxwell Montes, a mountain over 12 kilometers high; craters; arachnoids; and coronae. (It is important to note that the vertical scale of these images has been exaggerated to enhance the differences in elevation.) Other animation shows how the entire planet would appear if it were stripped of its cloud cover. Several astronomers describe the geologic features of Venus, its interior structure, the apparent absence of plate tectonics, and specific similarities and differences between Venus and Earth.

The Venusian atmosphere is the next topic covered in the video program. Several scientists supply information on the planet's atmospheric pressure, the constituent gases, and surface temperature. A planetary scientist and planetary meteorologist explain how conditions on Venus have produced a runaway greenhouse effect and compare the conditions on Venus with those on Earth.

The last half of the video program explores Mars, the most Earthlike planet in our solar system. Astronomers describe the early ideas about "mythic" Mars that existed up until the 1930s: about areas thought to be vegetation that changed with seasons and about polar ice caps. Although these features were seen through optical telescopes, it was not until the Mariner and Viking missions to Mars that planetary scientists were able to understand the features they had observed on the planet.

The video program includes the first photographs from space of another world, taken by the Mariner 4 spacecraft in its flyby of Mars in 1965. These black-and-white pictures showed a cratered, barren world, quite different from that imagined by science-fiction fans. Later missions supplied data that were used to create computer-generated three-dimensional images of the Valles Marineris (the Grand Canyon of Mars) and of Olympus Mons (a shield volcano). These images illustrate the geologic characteristics of Mars.

The segment on Mars next examines the evidence for liquid a history of water on the planet and possible explanations of why that water is no longer present. Spacecraft photos show the dry streambeds on the Martian surface, and computer animation illustrates water-filled streambeds drying up. Planetary scientists describe current thinking about the composition and behavior of the polar ice caps and about the large dust storms that produce terraces around the Martian poles.

The program concludes with Jeff Goldstein, an astrophysicist at the National Air and Space Museum, explaining that the laws of physics, although the same throughout our solar system, manifest themselves differently on these other worlds than on Earth. He notes that study of the physics on other worlds increases our understanding of the physical laws that govern phenomena on Earth.

As you watch the video program, consider the following questions:

1. What are the difficulties in observing the planet Venus?

2. How has radar astronomy allowed us to "see" the surface of Venus?

3. What explanation is given for the retrograde rotation of Venus?

4. Why doesn't Venus exhibit evidence of continental drift and plate tectonics?

5. What is the greenhouse effect and why does it cause such high temperatures on the surface of Venus?

6. What were the early Earth-based telescopic observations of Mars and what conclusions were drawn based on those observations?

7. What is a shield volcano and why are the ones on Mars so much higher than terrestrial volcanoes?

8. What is the evidence that suggests that Mars had large amounts of liquid water flowing on its surface some time in its past and where is the water on Mars now?

Review Activities

Matching

Match the terms listed below with the definitions that follow. Check your answer with the Answer Key and review any terms you missed.

_f_ 1. Magellan

_L_ 2. Maxwell Montes

_G_ 3. Ishtar Terra

_b_ 4. shield volcanoes

_i_ 5. rolling plains

_o_ 6 coronae

_e_ 7. arachnoids

_m_ 8. blob tectonics

_c_ 9. pancake domes

_H_ 10. greenhouse effect

_a_ 11. Olympus Mons

_n_ 12 Valles Marineris

_K_ 13. Tharsis region

_v_ 14. carbon dioxide

_d_ 15. Phobos

a. largest volcano on Mars

b. wide, low-profile volcanic cone produced by highly liquid lava

c. flat circular shapes approximately 65 kilometers across and 1 kilometer high that formed when viscous lava leaked through cracks on the surface of Venus

d. larger of the two moons of Mars

e. on Venus, geological patterns of circular and radial faults resembling spider webs

f. spacecraft that orbited Venus and mapped its surface using radar

g. highland region on Venus larger than the United States and containing one of the largest volcanoes on the planet

h. the process by which a carbon dioxide atmosphere traps heat and raises the temperature of a planetary surface

i. the relatively smooth lowlands that make up the largest proportion of the surface of Venus

j. most abundant gas in the atmosphere of Mars and Venus

k. on Mars, area in which several large volcanoes are located, including the planet's largest

l. one of largest volcanoes on the surface of Venus

m. process that describes how large masses of molten magma rise up below the crust of Venus and produce coronae, arachnoids, and other bubble/pancake-like features

n. a 2,500-mile long canyon on Mars

o. large circular features on the surface of Venus that appear to be slightly domed, apparently caused by rising plumes of molten rock below the crust

Completion

Fill each blank with the most appropriate term from the list for that paragraph. A term may be used once, more than once, or not at all. Check your answers with the Answer Key and review when necessary. If a question requires two or more answers in succession, they may be in any order, unless the question indicates otherwise.

1. In comparison with the atmosphere of Venus, Earth's atmosphere is many times _More_ dense. The Venusian atmosphere is also composed mostly of the gas _Ca Dioxin_ and, as a result, the surface temperature is hot enough to melt lead. The high surface temperature is produced by a phenomenon called the _Green House_. Because the temperatures on Venus are so high, this phenomenon is often referred to as a _Run away Greenhouse effect_.

Although the atmosphere of Venus is _Thick_ and acts as a _Blanket_ it still allows the sun's _heat_ to reach the surface of the planet. This energy is absorbed by the _Surface_ and reradiated away in the form of _Infrared_ energy. Because of the content of the atmosphere, much of the energy does not escape into space, but is trapped by the _Carbon Dioxide_ in the atmosphere. As a result, the surface of the planet _increases_ in temperature to extremely _High_ levels.

blanket	lead
carbon dioxide	less
decreases	low
escape	more
extreme	radiation
greenhouse effect	runaway greenhouse effect
high	surface
increases	thick
infrared	trapped

2. The major constituents of Earth's atmosphere are _Oxygen_ and _Nitrogen_. However, Earth's atmosphere also has _CO₂_, and, as a result, Earth also experiences a _Green House_, which increases the surface temperature. The heating that occurs on Earth, however, is much _less_ than it is on _Venus_. The principal reason for this difference is because Earth has _liquid water_ on its surface.

The early atmospheres of both Venus and Earth probably contained abundant quantities of the gases _CO₂_ and _water vapor_. However, because Earth is _farther_ from the sun, its early temperature was _lower_, which allowed the _water vapor_ in the atmosphere to condense into _liquid water_. The atmospheric _CO₂_ began to _dissolve_ into the _oceans_, forming mineral deposits and _carbonate rocks_. In other words, much of the _CO₂_ that once was in the atmosphere of Earth is now in the shells of shellfish and at the bottom of the oceans in the form of _limestone_ deposits.

carbon dioxide	liquid water
carbonate rocks	lower
dissolve	Mars
extreme	nearer
farther	nitrogen
greenhouse effect	oceans
increase	oxygen
less	Venus
limestone	water vapor

3. Our detailed knowledge of the surface of Mars comes from the images sent back to us from the _____ and _____ spacecraft that explored Mars in the _____. These spacecraft revealed Mars to be a geologically interesting place. The terrain in the southern hemisphere is heavily _____, which gives the landscape the appearance of the _____. The northern hemisphere has fewer _____, and they are less _____ than those in the southern hemisphere and therefore _____. Much of the terrain in the northern hemisphere is due to extensive _____ that has buried the original surface.

The volcanoes on Mars that have produced extensive _____ are of the _____ type, similar to the volcanoes of _____. However, the volcanoes on Mars are larger than the volcanoes on Earth. The size difference in volcanoes results from the weaker gravity on Mars and from a _____ Martian crust. This characteristic of the Martian crust is due to the fact that Mars is _____ than Earth and therefore _____ more quickly, allowing for a _____ crust to develop. This crust enabled the volcanoes to _____ in size until they are much _____.

However interesting the observed geologic activity on Mars was, planetary scientists were especially interested to find _____. These features indicate that, at some time in the Martian past, there must have been _____ on the surface. In fact, some observed features indicate that there may have been major _____ on the Martian landscape. Today, however, Mars has no _____ on its surface. However, there may be large quantities of _____ under the surface and at the Martian _____. The present lack of _____ on the surface demonstrates that the Martian climate has changed. The atmosphere of Mars is _____ than it was in the past and can no longer maintain a high enough _____ to keep _____ in its _____ form. The change in atmosphere probably results from the fact that Mars is _____ than Earth and its _____ cannot hold onto the gases to maintain a more _____ atmosphere.

1970s	gravity	poles
1980s	grow	pressure
Alaska	Hawaii	shield
cooled	ice	smaller
cratered	larger	thicker
craters	lava flows	thinner
dense	liquid	Viking
dry streambeds	liquid water	volcanic activity
eroded	Mariner	water
flooding	moon	younger

Self-Test

Select the one best answer for each question. Check your answers with the Answer Key and review when necessary.

1. The Pioneer spacecraft discovered that roughly 70 percent of the surface of Venus consists of

   a. volcanoes.

   b. fault zones.

   c. craters.

   d. rolling plains.

2. Spacecraft images of Venus offer little evidence for

   a. volcanic activity.

   b. major plate motion.

   c. the existence of mountains on the surface.

   d. continent-sized land masses.

3. Analysis of the surface by spacecraft that have landed on Venus indicates that the surface material consists of

   a. basalt.

   b. water.

   c. liquid magma.

   d. silicate carbonates.

4. The Magellan and Venera spacecraft located similar features, called coronae and arachnoids, on the surface of Venus. These features are

   a. large craters produced by asteroid impacts.

   b. volcanic cones.

   c. circular features probably caused by rising columns of magma under the crust.

   d. eroded shield volcanoes.

5. The major component of the atmosphere on Venus is

   a. nitrogen.

   b. oxygen.

   c. sulfuric acid.

   d. carbon dioxide.

6. The high temperatures on the surface of Venus are due to

   a. the planet's slow rotation rate.

   b. volcanic activity on the surface.

   c. a "runaway" greenhouse effect.

   d. reflective properties of the atmosphere.

7. The atmosphere of Venus contains much more carbon dioxide than the atmosphere of Earth. This difference is probably the result of

   a. the presence of more carbon dioxide in the primeval atmosphere of Venus.

   b. the slightly warmer temperatures on Venus that prevented the accumulation of great amounts of liquid water, thus preventing the dissolution of carbon dioxide.

   c. most of the carbon dioxide in Earth's atmosphere escaping into space.

   d. carbon dioxide still being produced by volcanoes on Venus.

8. Earth's crust is broken into large plates that slide over Earth's mantle. On Venus,

   a. similar plate activity has been observed.

   b. the plates are much smaller and move more quickly.

   c. there is no evidence of any kind of plate motion.

   d. there is little evidence that plate tectonics played a major role in shaping the surface of the planet.

9. Evidence of liquid water existing in the Martian past is found in the discovery of

   a. dry streambeds.

   b. oceans of liquid water currently on the surface.

   c. a dense atmosphere.

   d. volcanic activity.

10. Evidence of extensive volcanic activity on Mars is found in the

   a. highly cratered southern hemisphere of Mars.

   b. Tharsis region in the northern hemisphere where there are numerous faults and lava flows.

   c. polar ice caps where spectrum analysis shows evidence of basalt rocks.

   d. southern hemisphere where volcanoes such as Olympus Mons are located.

11. The polar caps of Mars consist of

   a. water ice.

   b. solid methane and carbon dioxide.

   c. solid nitrogen ice and liquid water.

   d. carbon dioxide and frozen water.

12. In the polar ice caps on Mars, water

   a. melts in the spring.

   b. never melts.

   c. evaporates in the summer.

   d. is not present.

13. The atmosphere of Mars consists primarily of

   a. carbon dioxide.

   b. oxygen.

   c. nitrogen.

   d. water vapor.

14. The density of the atmosphere on Mars is about

    a. equal to Earth's atmosphere.

    b. 50 percent of Earth's atmosphere.

    c. 10 percent of Earth's atmosphere.

    d. 1 percent of Earth's atmosphere.

15. The moons of Mars are

    a. Phobos and Deimos.

    b. Io and Ganymede.

    c. Hellas and Ishtar Terra.

    d. Maxwell Montes and Elysium.

16. The moons of Mars can be briefly described as

    a. smooth and nearly spherical in shape.

    b. being lightly cratered.

    c. irregular in shape with numerous surface craters.

    d. nearly identical in surface features.

Using What You've Learned

1. Locate the planets Venus and Mars on the SC1 Star Chart. Use the longitudes given in the table in Lesson 21. Choose a date near the current one and locate the sun, then determine if Venus or Mars can be seen in the nighttime sky. If you can see one of the planets, go out and observe it.

2. Compare the characteristics of the atmosphere and surface features of Venus with those of Earth. Do you feel that, in some respects, Earth is becoming more like Venus? Give examples.

3. Develop a diagram, including Earth, the sun, and Venus, that shows the cycle of the phases of Venus as viewed from Earth. If Venus is currently visible, use a pair of binoculars or a small telescope to observe the planet and determine where on your diagram Venus must be in order for you to observe its current phase.

4. Try to observe Mars with a small telescope. Draw the surface features that you can see. Try to identify these features on a detailed map of Mars.

5. Read a science fiction novel about the planet Mars. For example, read *Princess of Mars* by Edgar Rice Burroughs (1912). Compare how Mars is described in this book with what we know about it today.

6. Research the dimensions of Olympus Mons and Mauna Loa, and of Valles Marineris and the Grand Canyon. Construct a scale model of these surface features and compare their sizes.

Answer Key

Matching

1.	f	9.	c
2.	l	10.	h
3.	g	11.	a
4.	b	12.	n
5.	i	13.	k
6.	o	14.	j
7.	e	15.	d
8.	m		

Completion

1. more, carbon dioxide, greenhouse effect, runaway greenhouse effect

   thick, blanket, radiation, surface, infrared, carbon dioxide, increases, high

2. oxygen, nitrogen, carbon dioxide, greenhouse effect, less, Venus, liquid water

   carbon dioxide, water vapor, farther, lower, water vapor, liquid water, carbon dioxide, dissolve, oceans, carbonate rocks, carbon dioxide, limestone

3. Mariner, Viking, 1970s, cratered, moon, craters, eroded, younger, volcanic activity

   lava flows, shield, Hawaii, thicker, smaller, cooled, thicker, grow, larger

   dry streambeds, liquid water, flooding, liquid water, ice, poles, water, thinner, pressure, water, liquid, smaller, gravity, dense

Self-Test

(Page numbers refer to the textbook.)

1. d   (Objective 1; page 386)
2. b   (Objective 1; page 390; video program)
3. a   (Objective 1; page 390)
4. c   (Objective 1; page 390; video program)
5. d   (Objective 2; page 391; video program)
6. c   (Objective 2; page 391; video program)
7. b   (Objective 3; page 391)
8. d   (Objective 3; page 390; video program)
9. a   (Objective 4; page 397; video program)
10. b   (Objective 4; pages 395–397)
11. d   (Objective 5; pages 399–400)
12. b   (Objective 5; page 399)
13. a   (Objective 6; page 401)
14. d   (Objective 6; page 401)
15. a   (Objective 7; page 402)
16. c   (Objective 7; page 402)

L E S S O N
23

Jupiter and Saturn

Assignments

For the most effective study of this lesson, we suggest that you complete the assignments in the sequence listed below.

Before Viewing the Video Program

- Read the Overview and Learning Objectives for this lesson. Use the Learning Objectives to guide your reading, viewing, and thinking.

- Read Chapter 18, "Worlds of the Outer Solar System," sections 18-1 and 18-2, pages 407–423, in the textbook.

- Read the Viewing Notes in this lesson.

View "Jupiter and Saturn" Video Program

After Viewing the Video Program

- Briefly note your answers to questions listed at the end of the Viewing Notes.

- Review all reading assignments for this lesson, especially the first five paragraphs of the textbook Chapter 18 summary on pages 434–435 and the Viewing Notes in this lesson.

- Write brief answers to review questions 1 through 4 at the end of textbook Chapter 18 to be certain you understand the text material.

- Complete the Review Activities in this guide to reinforce your understanding of important terms and concepts. Check your answers with the Answer Key at the end of this lesson and review when necessary.

- Take the Self-Test in this guide to measure your achievement of the Objectives. Check your answers with the Answer Key and review when necessary.

- Complete the Using What You've Learned activities and any other activities and projects assigned by your instructor.

Overview

This lesson begins the tour of the cold regions of the outer solar systems, and the first worlds visited are Jupiter and Saturn, the two largest planets in the solar system. Both are gas/liquid giants and both have a family of orbiting moons and rings encircling them. Jupiter is five times the Earth-sun distance, and Saturn is ten times the Earth-sun distance. These planets, at such great distances from the sun, are not made of the familiar materials found on Earth. Hydrogen and helium, so rare on Earth, are the major components of these worlds. Not only are these elements abundant on these planets, they are in forms that are difficult, if not impossible, to duplicate in Earth laboratories. The environment of the outer planets is extremely unlike that of the inner planets. The temperatures at which the outer planets formed were much lower and the results are strikingly different. Even the moons of these worlds are also unusual. Instead of being made of the same materials as the planets they orbit, such as is the case with our moon, most of these satellites are composed of more ice than rock. One of the moons has an atmosphere. In fact, at increasingly greater distances from the sun, the moons of the outer planets tend to contain ever-higher percentages of ice.

The most massive object in the solar system, except for the sun, is Jupiter. Visible from Earth as the fourth-brightest object in the sky, Jupiter has retained its royal position in mythology and in current planetary research. The planet is an environment of extremes that provides astronomers with an opportunity to observe phenomena that cannot be duplicated on Earth.

Jupiter is surrounded by a satellite system. In orbit around the planet are more than a dozen moons, four of which have been photographed by the Voyager spacecraft. Those images reveal surfaces and environments that have not been observed elsewhere in the solar system: active volcanoes, smooth surfaces that have cracked and refrozen, cratered terrain, grooves, and multiple ringed craters. With its many unique features, the Jupiter system provides many challenges for planetary scientists.

Saturn, the other giant of the solar system, is almost as large as Jupiter. Unlike Jupiter, Saturn has an extensive system of rings. Although astronomers have known for quite some time that the rings were not solid, but made of small, icy particles, the arrangement of those particles was totally unexpected. The Voyager images showed such an intricate arrangement of ring particles that scientists had to explore new applications of Newton's law of universal gravitation.

Like Jupiter, Saturn has a satellite system in orbit around it. However, these moons are different from the moons of Jupiter and those in the inner solar system. They contain more ice than rock. One moon even has an atmosphere that is denser than Earth's. As astronomers learn more about Jupiter, Saturn, and their moons, they have realized that they must consider additional factors must be considered when explaining the nature and formation of the diverse objects in the solar system. Not only must the material of which the planets and moons are composed be accounted for, their distance from the sun needs to be included in the picture.

What astronomers learn from studying these immense, cold worlds is one more contribution to our understanding of the content, origin, and evolution of the solar system.

Learning Objectives

When you have completed all assignments for this lesson, you should be able to:

1. Describe the size, shape, rotation, and orbital characteristics of Jupiter and Saturn and describe their appearance as seen through a telescope. TEXTBOOK PAGES 407–413 and 418–419; VIDEO PROGRAM.

2. Recount the discoveries made by the Voyager spacecraft that explored the Jupiter and Saturn systems. TEXTBOOK PAGES 413–422.

3. Identify the Galilean satellites in orbit around Jupiter, provide details of their physical characteristics, and explain the origin of the identified features. TEXTBOOK PAGES 414–418; VIDEO PROGRAM.

4. Compare and contrast the ring systems of Jupiter and Saturn and explain the origin of these ring systems. TEXTBOOK PAGES 413–414 and 419–423; VIDEO PROGRAM.

5. Describe the current models of the atmosphere and interior of Jupiter and Saturn and explain the similarities and differences. TEXTBOOK PAGES 408–413 and 418–419.

6. Identify the major moons in orbit around Saturn, describe their physical characteristics, and explain the origin of these features. TEXTBOOK PAGE 423; VIDEO PROGRAM.

7. Describe in detail the characteristics of the extensive ring system around Saturn and explain its structure. TEXTBOOK PAGES 419–423; VIDEO PROGRAM.

8. Compare and contrast the characteristics and physical appearance of the large storms (spots) on Jupiter and Saturn. TEXTBOOK PAGES 411–413 and 418; VIDEO PROGRAM.

Viewing Notes

The two giants of the outer solar system—Jupiter and Saturn—are featured in this video program.

The first half of the program covers Jupiter and includes images of that planet taken on the Voyager missions, which explored the outer solar system and provided astronomers with vast quantities of information. In the program, astronomers describe what is currently known about Jupiter's features, including its unusual liquid composition of principally hydrogen and helium under tremendous pressure and its colorful atmosphere, full of cloud patterns and large areas of violent storms, and marked with the Great Red Spot. Images of Jupiter taken by spacecraft illustrate many of the features described in the program. This segment also includes information on Jupiter's rapid rotation and magnetic field.

Jupiter, along with the other Jovian planets, has both moons and rings. The Voyager explorations have provided excellent images of those objects, and astronomers explain what they have learned about moons Io, Europa, Ganymede, and Callisto and about the small particles that form the rings.

Saturn is featured in the second half of the video program. Astronomers point out the many similarities between the atmospheres and interiors of Saturn and Jupiter as well as noting some significant differences. Particular attention is focused on Saturn's spectacular concentric ring system, including descriptions of the system's structure and theories about the origin of the rings.

The segment on Saturn also includes information on Saturn's moons, including Titan with its cold atmosphere (rich in nitrogen but also having methane) and what has been learned from atmospheric studies and radar observations of Titan. Astronomers briefly describe Enceladus and Mimas, two of Saturn's other moons.

As you watch the video program, consider the following questions.

1. What are the primary components of Jupiter and Saturn? What are the gases?

2. How does the rapid rotation of Jupiter affect its cloud patterns?

3. What is the Great Red Spot?

4. What evidence, observed from Earth, revealed that Jupiter had a magnetic field? How is the magnetic field produced?

5. How do the characteristics of the Galilean satellites depend on their distance from Jupiter?

6. How do scientists explain the volcanic activity on Io?

7. What is the source of particles for Jupiter's ring?

8. How is the physical appearance of Saturn's clouds different from that of Jupiter's?

9. What is the composition and structure of Saturn's rings?

10. How do astronomers explain the intricate structure of Saturn's rings?

11. What are the similarities and differences between the moons of Jupiter and Saturn?

Review Activities

Matching

Match the terms listed below with the definitions that follow. Check your answers with the Answer Key and review any terms you missed.

_C_ 1. magnetosphere

_e_ 2. belt-zone circulation

_b_ 3. Roche limit

_G_ 4. grooved terrain

_a_ 5. tidal heating

_f_ 6. oblateness

_h_ 7. shepherd satellite

_d_ 8. spokes

_i_ 9. forward scattering

a. heating of a planet or satellite because of friction caused by tidal forces

b. minimum distance between a planet and a satellite that holds itself together by its own gravity

c. volume of space around a planet within which the motion of charged particles is dominated by the planetary magnetic field rather than by the solar wind

d. radial features in the rings of Saturn

e. atmospheric behavior typical of Jovian planets

f. flattening of a spherical body

g. regions of surface of Ganymede produced by repeated fracture and refreezing of the icy crust

h. object that, by its gravitational field, confines particles to a planetary ring

i. optical property of finely divided particles to preferentially direct light in the original direction of the light's travel

Completion

Fill each blank with the most appropriate term from the list for that paragraph. A term may be used once, more than once, or not at all. Check your answers with the Answer Key and review when necessary. If a question requires two or more answers in succession, they may be in any order, unless the question indicates otherwise.

1. The four largest moons of Jupiter were discovered by _____ and are referred to as the _____ satellites. The moon nearest Jupiter is _____. It is quite different from the other moons because is has virtually no _____, unlike the abundant number found on our own moon. The fact that this satellite has no _____ indicates that its surface is very young. Images taken by the _____ spacecraft revealed the reason. This moon is covered with active _____. The cause of this activity appears to be this moon's _____ orbit. The gravitational influence of Jupiter varies as the moon's distance from the planet varies. The moon is flexed by _____, which generate tremendous amounts of _____ by internal friction.

   The other three moons, or satellites, are equally as interesting. The second one out is _____. It also has no _____ on its surface, indicating that the surface is also very _____ and active. Its density is higher than those of the outer two moons, indicating a proportionally larger _____ core. This core is believed to be covered by a _____ mantle and a thick, _____ crust. The largest of Jupiter's moons is _____, the third of the _____ satellites. This moon is _____ and is also marked by _____, which appears to be breaks in the crust. These features formed _____ the period of cratering and are therefore relatively _____. The last of the four large satellites of Jupiter is _____. It is the most _____ of all the other moons and therefore has the _____ surface. Its density is less than half that of Earth's density. This low density is best explained by a composition that is a fifty-fifty mixture of _____ and _____.

after	Ganymede	oldest
before	gas	rock
Callisto	grooved terrain	rocky
cratered	heat	tidal forces
craters	ice	Titan
elliptical	icy	volcanoes
Europa	Io	Voyager
Galilean	liquid water	young
Galileo		

2. The most impressive feature of the planet Saturn is its spectacular ring system. The first to observe the rings was _____, although he did not know what he was actually viewing. From Earth, the rings were discovered to contain billions of small _____ and a few _____. The largest break in the rings is called the _____ division. Details of the rings were revealed in images obtained by the _____ spacecraft. These images showed that the rings are very complex, consisting of thousands of narrow _____. In addition, new rings were discovered. One ring in particular was very twisted and braided. The structure of this ring is caused by _____ small _____. As a result of the _____ influences of these objects, the ring particles are kept in a very narrow band. These objects are called _____ satellites. Other gaps in the rings are caused by gravitational _____ between the ring particles and _____. For example, the largest break in the rings is the result of a tug of war between the ring particles and the moon _____. Remember, however, not all the breaks in the rings can be explained by this kind of gravitational tug of war, especially the hundreds of _____.

Cassini	moons	shepherding
Galileo	particles	three
gaps	resonances	two
gravitational	ringlets	Voyager
Mimas	satellites (moons)	

Self-Test

Select the one best answer for each question. Check your answers with the Answer Key and review when necessary.

1. The largest planet in the solar system is
   a. Jupiter.
   b. Saturn.
   c. Uranus.
   d. Neptune.

2. The oblateness of Jupiter and Saturn is due to their
   a. ring system.
   b. rapid rotation.
   c. numerous satellites.
   d. gravitational resonances.

3. As viewed through a telescope, Jupiter appears as a
   a. spherical and heavily cratered object.
   b. bland and colorless disk.
   c. disk composed of various colored belts and bands.
   d. highly elliptical and bland disk.

4. Saturn, as viewed through a telescope, appears as a
   a. spherical and heavily cratered object.
   b. bland and nearly featureless disk.
   c. disk composed of brightly colored belts and bands.
   d. highly elliptical and bright blue disk.

5. The Voyager spacecraft discovered volcanoes on the moon
   a. Callisto.
   b. Titan.
   c. Ganymede.
   d. Io.

6. The Voyager 1 spacecraft discovered a ring system around
   a. Io.
   b. Saturn.
   c. Jupiter.
   d. Ganymede.

7. The grooved terrain on Ganymede is the result of
   a. volcanic activity on the surface.
   b. repeated breaking, melting, and refreezing of the icy crust.
   c. asteroid impacts.
   d. cometary impacts.

8. The surface of Io is covered with
   a. ice.
   b. sulfur ash.
   c. nitrogen.
   d. methane.

9. The ring particles around Jupiter are very small, as evidenced by the observation that light
   a. produces a rainbow of colors.
   b. reflects off the particles in the backward direction.
   c. scatters in the forward direction.
   d. is absorbed by the ring particles.

10. The rings of Saturn lie within the Roche limit, thereby
    a. preventing the formation of a large satellite.
    b. producing the Cassini division.
    c. allowing shepherding satellites to form and create gravitational resonances.
    d. preventing the development of ringlets.

11. The core of Jupiter is composed of
   a. rock and ice.
   b. liquid and gaseous hydrogen.
   c. liquid helium.
   d. heavy elements.

12. The belts and bands of Jupiter are the result of atmospheric features that have been
   a. brought up by convection currents from deep within Jupiter's interior.
   b. refracted through the hydrogen and helium atmosphere.
   c. distorted by cometary impacts.
   d. stretched and elongated as a result of Jupiter's rapid rotation.

13. The atmosphere of Titan is composed primarily of
   a. oxygen.
   b. nitrogen.
   c. methane.
   d. There is no atmosphere on Titan.

14. Some regions of the moon Enceladus are a thousand times less cratered than other regions. These lightly cratered regions
   a. missed the cratering epoch in the evolution of the solar system.
   b. probably have craters so small they were not observed by Voyager.
   c. are very young.
   d. are very old.

15. The Cassini division is a large gap in the rings of Saturn. This gap is most likely produced by
   a. the reflection of sunlight off dark material.
   b. resonances between ring particles and the moon Mimas.
   c. shepherding satellites.
   d. the absorption of sunlight by dark material.

electrostatic

16. Radially oriented features in the rings of Saturn, called spokes, are produced by

    a. shadows caused by dust particles pulled out of the ring plane.

    b. dark material emanating from the inner rings.

    c. collisions between ring particles.

    d. electromagnetic forces between the dust particles and the shepherding satellites.

17. The extended life of Jupiter's Great Red Spot is probably because the Great Red Spot

    a. originates from the core of Jupiter.

    b. is related to the position of the moon Io.

    c. lies continually over a liquid "surface."

    d. generates its own thermal energy.

18. The atmosphere of Saturn shows numerous and colorful "spots," similar to those on Jupiter. These spots result from

    a. thermal radiation.

    b. extreme temperatures, both high and low.

    c. the gravitational pull of Titan.

    d. Saturn does not display numerous and colorful spots, because of its very low cloud temperatures.

Using What You've Learned

1. Refer to an astronomy magazine that provides observing information for Jupiter and the Galilean satellites of Jupiter. With a small telescope or good binoculars, identify the four Galilean satellites and sketch their positions. Keep observing these moons for a period of several hours over the course of an evening. Which moons appear to "disappear" as they pass in front of the planet? Which moons "disappear" as they pass behind the planet?

2 Consult a current astronomy magazine to locate Saturn's position and then go out at night and try to observe the planet. Good binoculars or a small telescope should also permit you to observe Saturn's rings and its large moon, Titan. Make a sketch of your observations. Try to include Cassini's division in your sketch. (You may also use the coordinates in Lesson 21 in this guide and the SC1 Star Chart.)

3. With the aid of a small telescope with an aperture of at least 7 centimeters, (3 inches), or very good binoculars, observe the large moon, Titan, over a period of a month. Initially, it may be difficult to distinguish Titan from the background stars that appear near Saturn. However, by sketching Saturn on successive nights and locating the nearby stars as seen in the telescope, and then comparing the sketches, you will observe two objects—Saturn and Titan—appearing to move together relative to the stationary background stars. Trace their paths relative to the background stars and relative to each other. From your sketch, you should be able to estimate the period of Titan's revolution around Saturn. How does this period compare with the revolutions of the Galilean satellites and why is Titan's period different?

Answer Key

Matching

1. c
2. e
3. b
4. g
5. a

6. f
7. h
8. d
9. i

Completion

1. Galileo, Galilean, Io, craters, craters, Voyager, volcanoes, elliptical, tidal forces, heat

   Europa, craters, young, rocky, liquid water, icy, Ganymede, Galilean, cratered, grooved terrain, icy, after, young, Callisto, cratered, oldest, rock, ice

2. Galileo, particles, gaps, Cassini, Voyager, ringlets, two, moons, gravitational, shepherding, resonances, satellites (moons), Mimas, ringlets

Self-Test

(Page numbers refer to the textbook.)

1. a (Objective 1; page 408; video program)
2. b (Objective 1; page 418)
3. c (Objective 1; page 411)
4. b (Objective 1; pages 418–419)
5. d (Objective 2; page 417)
6. c (Objective 2; page 413)
7. b (Objective 3; page 416)
8. a (Objective 3; page 416–417)
9. c (Objective 4; page 413)
10. a (Objective 4; page 422)
11. d (Objective 5; page 408)
12. d (Objective 5; page 411)
13. b (Objective 6; page 423)
14. c (Objective 6; page 423)
15. b (Objective 7; page 422)
16. a (Objective 7; page 422)
17. c (Objective 8; video program)
18. d (Objective 8; video program)

Jupiter mass

CREI

L E S S O N
24

Uranus, Neptune, and Pluto

Assignments

For the most effective study of this lesson, we suggest that you complete the assignments in the sequence listed below

Before Viewing the Video Program

- Read the Overview and Learning Objectives for this lesson. Use the Learning Objectives to guide your reading, viewing, and thinking.

- Read Chapter 18, "Worlds of the Outer Solar System," sections 18-3 through 18-5, pages 424–434, in the textbook.

- Read the Viewing Notes in this lesson.

View "Uranus, Neptune, and Pluto" Video Program

After Viewing The Video Program

- Briefly note your answers to questions listed at the end of the Viewing Notes.

- Review all reading assignments for this lesson, especially the textbook Chapter 18 summary on pages 434–435 and the Viewing Notes in this lesson.

- Write brief answers to review questions 6–10 at the end of textbook Chapter 18 to be certain you understand the text material.

- Complete the Review Activities in this guide to reinforce your understanding of important terms and concepts. Check your answers with the Answer Key at the end of this lesson and review when necessary.

- Take the Self-Test in this guide to measure your achievement of the Objectives. Check your answers with the Answer Key and review when necessary.

- Complete the Using What You've Learned activities and any other activities and projects assigned by your instructor.

Overview

This lesson on Uranus, Neptune, and Pluto takes us to the edge of our solar system and completes our study of the planets. Much of what astronomers know about two of these most distant planets—Uranus and Neptune—was learned quite recently from data and images sent back by Voyager spacecraft.

Of course, the information provided by the Voyager missions had been valuable in studying the other planets, but it was exceptionally valuable for learning more about Uranus and Neptune, because the great distances to those planets had impeded their study with Earth-based telescopes.

Astronomers were surprised and challenged by much of the information sent back by Voyager: volcanoes on Io, ringlets around Saturn, shepherding moons. Information about Uranus and Neptune was even more surprising and challenging. The spacecraft saw dark rings around Uranus and, passing Neptune, Voyager sent back striking images of clouds and a dark spot and rings composed of dark material.

Even the moons around Uranus and Neptune surprised astronomers. One moon around Uranus appears to have been shattered and then reassembled, while others have regions on their surfaces that show very little cratering. The moon Triton, orbiting Neptune in nearly the coldest region of the solar system, is volcanically active. The question for astronomers, of course, was: Where does the heat come from? Even Pluto, the smallest planet in the solar system, and its moon Charon, raised questions for astronomers about the history and formation of the solar system. Astronomers are debating whether Pluto, Charon, and Triton are a new class of objects orbiting in the farthest reaches of the solar system, and if collisions with these kinds of objects caused the peculiar behavior of Uranus.

All the information about the planets and associated objects gathered and studied by astronomers is important in developing and refining theories on the formation, history, and evolution of the solar system. This knowledge is important, not only because knowledge is important for its own sake, but also because it helps to put planet Earth in perspective. Earth is a planet orbiting around a relatively common star. It exists in an environment with other planets and minor objects. By studying the entire solar system, astronomers can develop an understanding of how this environment affects the planet and this life on its surface.

Learning Objectives

When you have completed all assignments for this lesson, you should be able to:

1. Describe the orbital, rotational, and physical characteristics of Uranus, Neptune, and Pluto as observed from Earth. TEXTBOOK PAGES 424–425, 428–429, and 432–434; VIDEO PROGRAM.

2. Describe the events leading to the discoveries of Uranus, Neptune, and Pluto. VIDEO PROGRAM.

3. Compare and contrast the characteristics of the atmospheres of Uranus and Neptune as observed by the Voyager spacecraft. TEXTBOOK PAGES 424–425 and 428–429; VIDEO PROGRAM.

4. Describe the events leading to the discoveries of the ring systems around Uranus and Neptune and explain how astronomers deduced that rings existed around Uranus and Neptune. TEXTBOOK PAGES 425–427 and 429; VIDEO PROGRAM.

5. Describe the features of the ring systems of Uranus and Neptune and explain the origin of these features. TEXTBOOK PAGES 427–428, 429, and 432.

6. Describe the physical characteristics of the moons of Uranus and Neptune and explain the origins of these features. TEXTBOOK PAGES 425–427 and 429–432; VIDEO PROGRAM.

7. Compare and contrast the orbital characteristics of the satellite systems of Uranus, Neptune, and Pluto. TEXTBOOK PAGES 427–428 and 429–432.

8. Describe the known physical characteristics of Pluto and its moon Charon. TEXTBOOK PAGES 432–433; VIDEO PROGRAM.

Viewing Notes

As with the other video programs on the planets, "Uranus, Neptune, and Pluto" provides descriptions of each planet, accompanied by NASA and JPL photographs.

The segment on Uranus begins by relating the story of how William Herschel accidentally discovered the planet in 1781. Astronomers go on to describe the planet's composition, the tip in its orbital axis, its rotation, magnetic field, and atmosphere. The video program also supplies information about Uranus' dark rings and the planet's moons, including Miranda and theories about that moon's unusual history.

In discussing Neptune in the next segment, astronomers explain that planet's similarities to Uranus and note important differences. The atmosphere of Uranus is more active, and the planet has white clouds and a great dark spot. The video program also supplies information on Neptune's rings and Triton, one of its two major moons. Planetary scientists describe some of icy Triton's unusual features, including its retrograde orbit, rocky interior, nitrogen and methane atmosphere, relative absence of craters, and dark "smudges," which may be caused by active volcanoes or geysers.

The last part of the video program focuses on Pluto, the most distant planet in the solar system, and its moon, Charon. Pluto is also the most recently discovered planet, and the story of its discovery by Clyde Tombaugh in 1930 is related. Because of its great distance, little is known about Pluto, but astronomers explain how they have been able to determine the planet's mass and how spectroscopy has provided information on the planet's composition. Planetary scientists describe current theories about the origin of this unusual planet as well as the origin of similar icy bodies in the distant areas of the solar system. Finally, astronomers briefly discuss the possibility of discovering other planets beyond Pluto.

As you watch the video program, consider the following questions.

1. Why is the disk of Uranus so featureless?

2. How were the rings of Uranus discovered?

3. What constituent of the atmosphere of Neptune makes it appear blue?

4. What feature on Neptune is comparable to the Great Red Spot on Jupiter?

5. How was Pluto discovered?

6. How was its moon Charon discovered?

Review Activities

Matching

Match the terms listed below with the definitions that follow. Check your answers with the Answer Key and review any terms you missed.

___i___ 1. methane   ___c___ 6. ring arcs

___e___ 2. occultation   ___a___ 7. retrograde orbit

___f___ 3. organic polymer   ___b___ 8. Charon

___b___ 4. ovoids   ___d___ 9. ice dwarfs (Plutons)

___4___ 5. cyclonic circulations

a. backward motion of satellite as it orbits a planet

b. oval features found on Miranda

c. sections of complete rings, unique to planet Neptune

d. icy bodies that may have once populated the outer solar system

e. passage of one celestial body in front of another

f. complex carbon-based matter that is very dark and sticky

g. Pluto's moon

h. system of rapidly rotating winds over a vast area

i. gas that makes the atmospheres of Uranus and Neptune appear blue

Matching—Planet Review

Now that you have completed your study of the planets of our solar system, the following activity will help you review what you have learned about these worlds. Match each description or feature in the numbered list below with the planet or planets to which it applies from the list that follows. Check your answers with the Answer Key and review any planetary features you missed.

_b_ 1. largest number of known satellites

_a_ 2. famous red spot

_c_ 3. surface with the highest temperature

_d_ 4. frozen water and evidence of former rivers

_e_ 5. smallest mass

_a_ 6. largest mass

_i_ 7. farthest from Earth

_d f_ 8. polar caps

_a_ 9. strongest magnetic field

_f_ 10. third from the sun

abhi 11. rings

_c_ 12. most like Earth in size

_j_ 13. like Earth in surface conditions

_a_ 14. closest to the sun

_e_ 15. most recently discovered

_i_ 16. discovered by mathematical prediction

_b_ 17. associated with Cassini's division

_c_ 18. largest amount of carbon dioxide in its atmosphere

_b_ 19. lowest mean density

_e_ 20. orbited by Charon

a. Jupiter	e. Pluto	i. Neptune
b. Saturn	f. Earth	j. none
c. Venus	g. Mercury	
d. Mars	h. Uranus	

Completion

Fill each blank with the most appropriate term from the list for that paragraph. A term may be use once, more than once, or not at all. Check your answers with the Answer Key and review when necessary. If a question requires two or more answers in succession, they may be in any order, unless the question indicates otherwise.

1. The last two giant planets of the outer solar system have significant similarities and differences when compared to Jupiter and Saturn. Both Uranus and Neptune are composed mostly of _____ and _____ just as are Jupiter and Saturn. Uranus and Neptune, however, do not have sufficient densities or pressures to produce an interior of liquid _____ hydrogen. The interiors of these worlds are simply _____ hydrogen. The cloud patterns of Uranus and Neptune are also noticeably different from those of Jupiter and Saturn. Through a telescope, both Uranus and Neptune appear to be _____ and they do not have the colorful _____ and _____ so visible in the clouds of Jupiter. In fact, Uranus is practically _____. The apparent color of Uranus and Neptune is due to the presence of _____ in the atmosphere. This gas absorbs _____ and _____ light, thus producing the _____ coloring.

   When astronomers first saw the images of Neptune taken by Voyager, they were surprised. Because of Neptune's great distances and very low _____, they had expected the planet to appear _____ and _____. However, Neptune showed _____ detail than Uranus and even had a feature similar to the Great Red Spot on Jupiter. The color of this feature is _____ and it is referred to as the _____. Astronomers concluded that Neptune must be producing some internal _____, causing _____. Combined with the _____ rotation of the planet, _____ winds and _____ circulations are generated.

bands	high-speed
belts	hydrogen
bland	less
blue	liquid
blue-green	metallic
cold	methane
convection	more
cyclonic	rapid
featureless	red
Great Dark Spot	slow
heat	temperatures
helium	yellow

2. The rings of Uranus and Neptune are very much different from the rings of Saturn. The rings of Uranus were detected when the planet _____ a star. Dips in the _____ from the star appeared _____ of the planet and a _____ pattern of dips appeared _____ the planet. The rings of Uranus are very _____. The structure of these rings depends on the presence of tiny _____, which control the position of the ring _____. These objects are called _____. Unlike the rings of Saturn, the rings of Uranus appear _____. It has been hypothesized that this color of the rings is due to the presence of _____. These chemicals were produced when the frozen _____ in the ring particles was converted into this _____ matter by the radiation associated with the _____ of Uranus.

ahead	narrow
behind	occulted
black	organic polymers
broad	particles
dark	radiation belts
light	shepherding satellites
methane	symmetrical
moons	

3. Because Pluto's orbit is highly _____, that planet is sometimes _____ to the sun than is Neptune. The two planets will never collide, because the orbit of Pluto is _____ to the orbit of Neptune, and the two objects will never be at the same place in the solar system at the same time. Because of its great distance, very little else was known about Pluto until the discovery of a moon in orbit about the planet.

The discovery of this moon, called _____, enabled astronomers to calculate the _____ of Pluto. By analyzing these two objects as they would an _____ star system, astronomers were able to determine the _____ of both objects. The size of this moon is about _____ that of Pluto. Calculations of the density of Pluto and the moon revealed they are composed mostly of a mixture of _____ and _____.

There is a highly speculative explanation for the origin of Pluto. Because of the density and size calculations that have been made, Pluto and its moon seem very similar to the moon _____. This conclusion leads scientists to believe that there are other _____ bodies beyond the orbit of Pluto. These objects have been tentatively called _____. Some astronomers think that a near-collision of Uranus with these objects caused a severe tilt to the planet's axis of _____ and the moon Triton's orbit around Neptune to be _____.

Callisto	mass
Charon	moon
closer	one-half
diameters	one-third
eclipsing binary	retrograde
elliptical	rocks
ice	rotation
ice dwarfs	Triton
icy	Uranus
inclined	

Self-Test

Select the one best answer for each question. Check your answers with the Answer Key and review when necessary.

1. Which of the following planets has its axis of rotation tipped the most?
   a. Uranus
   b. Neptune
   c. Pluto
   d. Charon

2. Which of the following planets has the most elliptical orbit?
   a. Uranus
   b. Triton
   c. Neptune
   d. Pluto

3. The discovery of Neptune was made as the result of
   a. a study of planetary nebula.
   b. a search for comets.
   c. predictions following mathematical calculations.
   d. a search for near-Earth asteroids.

4. Percival Lowell believed that the irregularities in the motion of the planet Uranus were due to a planet beyond Neptune. His efforts and the efforts of others that followed led to the eventual discovery of
   a. Triton.
   b. Nereid.
   c. Pluto.
   d. Charon.

5. The very bland appearance of the atmosphere of Uranus is probably due to the
   a. absence of hydrogen and helium.
   b. extremely cold temperatures and the distance of Uranus from the sun.
   c. absence of methane in the atmosphere.
   d. presence of methane in the atmosphere.

6. Neptune's blue appearance is the result of the presence of

    a. hydrogen.

    b. helium.

    c. ammonia.

    d. methane.

7. The rings of Uranus were discovered during

    a. the Voyager spacecraft flyby.

    b. the occultation of a star by Uranus.

    c. an eclipse of Uranus by the moon Miranda.

    d. an occultation of Uranus by Saturn.

8. The rings of Neptune were discovered as the result of an occultation of a star by Neptune. Astronomers concluded that the rings were not complete because

    a. only one side of the planet yielded the expected dimming of the starlight.

    b. only half the rings were visible from Earth.

    c. theoretical calculations predicted these results.

    d. only half the rings were observed by the Voyager spacecraft.

9. The ring systems of Uranus and Neptune consist of very narrow rings. This narrowness is probably the result of

    a. random collisions that have driven particles out of the ring system.

    b. magnetic constrictions resulting from the interaction of the particles with the magnetic fields of the planets.

    c. shepherding satellites controlling the positions of the ring particles.

    d. solar radiation pressure.

10. The rings of Uranus are extremely dark. This appearance is probably due to the presence of

    a. volcanic ash.

    b. organic polymers.

    c. frozen methane.

    d. inorganic ammonia compounds.

11. The peculiar appearance of the moon Miranda is explained by
    a. gravitational influences produced by Uranus.
    b. tidal influences produced by the other nearby moons Umbriel and Ariel.
    c. collisions between Miranda and the ring particles.
    d. a large impact breaking Miranda into pieces, with the pieces then falling back together.

12. Dark smudges on the surface of Triton appear to be the result of
    a. organic polymers resulting from ammonia interacting with Neptune's magnetic field.
    b. impact debris from asteroid collisions.
    c. nitrogen volcanism.
    d. disintegration of ring particles falling on the planet.

13. Neptune's moon Triton
    a. is in a retrograde orbit.
    b. is in a perfectly circular orbit.
    c. orbits in the same direction as the other moon, Nereid.
    d. has a polar orbit.

14. Pluto and Charon are tidally locked, thereby
    a. preventing Charon from orbiting Pluto.
    b. causing Charon to orbit Pluto faster than it turns on its own axis.
    c. causing Charon to orbit Pluto slower than it turns on its own axis.
    d. causing Pluto and Charon to keep the same sides facing each other.

15. Pluto's moon Charon is approximately
    a. half the size of Pluto.
    b. twice the size of Pluto.
    c. one-fourth the size of Pluto.
    d. equal to the size of Pluto.

16. The diameters of both Pluto and Charon were determined

   a. by direct observation from Voyager.

   b. from the analysis of a series of mutual eclipses as one would study eclipsing binary star systems.

   c. by direct observation from Earth-based telescopes.

   d. by direct observation from the Hubble Space Telescope.

Using What You've Learned

1. Describe the seasons as they would occur on Uranus, and compare them to Earth's.

2. Outline the events that led to the discoveries of Uranus, Neptune, and Pluto. What role did serendipity play?

3. Explain why Pluto, at times, is not the most distant planet in the solar system.

Answer Key

Matching

1.	i	6.	c
2.	e	7.	a
3.	f	8.	g
4.	b	9.	d
5.	h		

Matching—Planet Review

1.	b	6.	a	11.	a, b, h, i	16.	i
2.	a	7.	i	12.	c	17.	b
3.	c	8.	d, f	13.	none	18.	c
4.	d	9.	a	14.	g	19.	b
5.	e	10.	f	15.	e	20.	e

Completion

1. hydrogen, helium, metallic, liquid, blue-green, belts, bands, featureless, methane, red, yellow, blue-green

   temperatures, bland, featureless, more, blue, Great Dark Spot, heat, convection, rapid, high-speed, cyclonic

2. occulted, light, ahead, symmetrical, behind, narrow, moons, particles, shepherding satellites, black, organic polymers, methane, dark, radiation belts.

3. elliptical, closer, inclined

   Charon, mass, eclipsing binary, diameters, one-half, ice, rocks

   Triton, icy, ice dwarfs, rotation, retrograde.

Self-Test

(Page numbers refer to the textbook.)

1. a (Objective 1; page 424)

2. d (Objective 1; page 432)

3. c (Objective 2; video program)

4. c (Objective 2; video program)

5. b (Objective 3; pages 424–425)

6. d (Objective 3, page 428)

7. b (Objective 4; page 425)

8. a (Objective 4; video program)

9. c (Objective 5; page 425)

10. b (Objective 5; page 425)

11. d (Objective 6; page 428)

12. c (Objective 6; page 431)

13. a (Objective 7; page 429)

14. d (Objective 7; page 434)

15. a (Objective 8; page 432)

16. b (Objective 8; page 432)

L E S S O N
25

Meteorites, Asteroids, and Comets

Assignments

For the most effective study of this lesson, we suggest that you complete the assignments in the sequence listed below.

Before Viewing the Video Program

- Read the Overview and Learning Objectives for this lesson. Use the Learning Objectives to guide your reading, viewing, and thinking.

- Read Chapter 19, "Meteorites, Asteroids, and Comets," pages 438–455, in the textbook.

- Read Background Notes 25A, "Kuiper Belt"; 25B, "Comet Halley—History"; 25C, "Comet Shoemaker–Levy 9"; and 25D, "Interplanetary Debris Impacts, following Lesson 26 in this guide."

- Read the Viewing Notes in this lesson.

View "Meteorites, Asteroids, and Comets" Video Program

After Viewing the Video Program

- Briefly note your answers to questions listed at the end of the Viewing Notes.

- Review all reading assignments for this lesson, especially the textbook Chapter 19 summary on pages 452–453, the Background Notes 25A–25D, and the Viewing Notes in this lesson.

- Write brief answers to all the review questions at the end of textbook Chapter 19 to be certain you understand the text material.

- Complete the Review Activities in this guide to reinforce your understanding of important terms and concepts. Check your answers with the Answer Key at the end of this lesson and review when necessary.

- Take the Self-Test in this guide to measure your achievement of the Objectives. Check your answers with the Answer Key and review when necessary.

- Complete any other activities and projects assigned by your instructor.

Overview

After learning about the planets of our solar system, you may have a tendency to think of the smaller objects in the solar system as "leftovers." These objects range in size from that of a very small moon down to dust particles. Some of these objects—comets and meteors—are responsible for startling visual effects. Others, such as asteroids, are invisible to the unaided eye and could be overlooked even if photographed through a telescope. Why does all this debris attract so much attention? There aren't any definite answers yet concerning the solar system's origins, but these objects and particles do contain intriguing clues on how the solar system was formed.

This lesson is concerned with the organization, classification, and analysis of the smallest pieces of the solar system. The "rocks" from outer space are correctly called meteorites and can be classified according to their appearance and composition. Astronomers now know that meteorites come from larger parent bodies: asteroids. Asteroids, sometimes called "minor planets," are the remaining planetesimals from which the planets and moons were made. Detailed study of meteorites helps scientists to learn about the history and early processes through which the asteroids have passed. The link between meteorites and asteroids allows scientists to look back into the earliest times of the solar system.

Unfortunately, most of the meteorites that arrive on Earth have been altered through heating or melting. Although these processes occurred in the earliest times of the solar system, meteorites, except for one type, do not reveal much about the original matter of the solar system.

In contrast to meteorites and asteroids, comets are the most pristine objects in the solar system. Made of ice and rock, any melting or heating would have caused comets to be obliterated within the 4.5 billion years of the solar system's existence. Since comets still exist and occasionally visit the vicinity of Earth, they must exist in a region of the solar system that is free from the high temperatures that would change their structures. In this outermost region of the solar system, scientists may one day find answers to the questions about the origin of the planets, their moons, and possibly life itself.

In the past, these "hairy stars" created much superstition and fear, and it was not until the nineteenth century when astronomers began to learn about comets and discredit some of the pseudoscientific myths that were associated with a comet's apparition.

The study of comets was Earth-based until the return of Comet Halley in 1986. At that time, spacecraft were launched to rendezvous with Halley. These spacecraft took photographs of the comet's nucleus, measured its characteristics, and sampled the contents of its tail. The encounter with Comet Halley provided new information about the shape, chemical composition, and the possibility that the crust of the comet is composed of carbon. Comets, once believed to be omens of death and destruction, could very well be the source of Earth's first oceans and material that eventually allowed life to form and evolve on this planet. The smallest pieces within the solar system may tell us the biggest story.

Learning Objectives

When you have completed all assignments for this lesson, you should be able to:

1. Identify the two broad categories of meteorites and the three types of stony meteorites. TEXTBOOK PAGES 439–440; VIDEO PROGRAM.

2. Describe the possible origins of the different types of meteorites. TEXTBOOK PAGES 440–442; VIDEO PROGRAM.

3. Describe the characteristics of meteors and meteor showers, and explain how they relate to cometary orbits. TEXTBOOK PAGES 440–442.

4. Describe the physical properties and orbital characteristics of asteroids. TEXTBOOK PAGES 442–443; VIDEO PROGRAM.

5. Explain theories about the origin of asteroids. TEXTBOOK PAGES 443–444; VIDEO PROGRAM.

6. Identify and describe the major parts of a comet and explain the effect of solar wind on the tail of a comet. TEXTBOOK PAGES 444–450; VIDEO PROGRAM.

7. Explain the significance of the Oort cloud and Kuiper belt theories to understanding the origin of comets. TEXTBOOK PAGES 450–451; BACKGROUND NOTES 25A; VIDEO PROGRAM.

8. Describe the significant historical observations of Comet Halley, including the 1910 apparition, and the important discoveries made by spacecraft during the 1986 apparition. TEXTBOOK PAGES 445–450; BACKGROUND NOTES 25B; VIDEO PROGRAM.

9. Explain the possible effects of an asteroid/comet collision with Earth and describe the impact of Comet Shoemaker-Levy 9 on Jupiter. TEXTBOOK PAGES 451–452; BACKGROUND NOTES 25C and 25D; VIDEO PROGRAM.

Viewing Notes

This video program concludes our exploration of the solar system with an examination of solar system "debris": meteorites, asteroids, and comets.

The segment on meteorites features astronomers explaining the composition of meteorites and various theories about their origin. The program also includes examples of the major types of meteorites.

Asteroids, thought to be the source of meteorites, are covered in the next segment. The video program provides information on the three classifications of asteroids—S types, C types, and M types—and the relationship of each classification to the different types of meteorites found on Earth's surface. Astronomers describe how the first asteroid, Ceres, was discovered in 1801. It was not until spacecraft carried telescopes into space that astronomers were able to learn about the surface features and shape of these small objects. In the program, you will see a photograph of the asteroid Gaspra, taken by the Galileo spacecraft in 1990.

Astronomers describe how most asteroids orbit in a belt between Mars and Jupiter and offer theories about the origin of asteroids. They also provide information on asteroids outside the main asteroid belt. Trojan asteroids precede and follow Jupiter in its orbit, and Apollo asteroids have elongated orbits that cross Earth's orbit. Some Apollo asteroids are in orbits that occasionally bring them close to Earth. Ultimately, these asteroids crash into a planet. The video program looks at a possible asteroid or comet impact crater near the Yucatán, in Mexico, and considers the theory that the aftereffects of the impact led to the extinction of the dinosaurs.

Comets, some of which produce spectacular celestial events as they arc through the sky, are the subject of the final segment. The segment includes information on Comet Halley, probably the most famous of all comets seen from Earth, and examines the lore surrounding some of the comet's past apparitions. In the program, you'll see photographs of Comet Halley taken by the spacecraft Giotto during the comet's most recent appearance in 1985-1986. These photographs supplied astronomers with valuable information about the comet's nucleus, and astronomers describe what is now known about a comet's head and tail and composition. They offer two principal theories—the Oort cloud theory and the Kuiper belt theory—about the origin of comets. This segment also includes footage of a most spectacular cometary event: the impact of Comet Shoemaker-Levy 9 on Jupiter in 1994. Staff from the Jet Propulsion Laboratory describe the collision of the comet's 20 fragments and offer some preliminary findings.

As you watch the video program, consider the following questions:

1. How do meteors, meteoroids, and meteorites differ?

2. What are the different types of meteorites and what kinds of information do these objects reveal?

3. What is an asteroid and where is the asteroid belt?

4. What are the different classes of asteroids and how are these classes related to the different types of meteorites?

5. What is a comet and what characteristics of a comet are visible through a telescope?

6. What are the parts and composition of a comet?

7. Where do comets originate?

8. What are possible effects of a cometary impact on Earth?

9. Why is Comet Halley so important and what was learned during its last visit to the inner solar system?

Review Activities

Match the terms below with the definitions that follow. Check your answers with the Answer Key and review any terms you missed.

Matching

_e_ 1. achondrite

_d_ 2. carbonaceous chondrite

_b_ 3. chondrite

_c_ 4. chondrule

_G_ 5. coma

_f_ 6. meteor shower

_H_ 7. Oort cloud

_a_ 8. Widmanstätten pattern

_j_ 9. Kuiper belt

_L_ 10. radiant point

_i_ 11. meteor

_M_ 12. meteorite

_K_ 13. meteoroid

a. bands in iron meteorites created by large crystals of nickel-iron alloys

b. a stony meteorite with chondrules

c. round, glassy body found in some stony meteorites

d. stony meteorite with both chondrules and volatiles

e. stony meteorite with composition similar to that of Earth's lavas

f. event that occurs when Earth passes near the orbit of a comet, probably caused by dust and debris released by a comet

g. glowing head of a comet

h. hypothetical source of comets

i. small bit of matter heated by friction to incandescent vapor as it falls into Earth's atmosphere

j. region beyond the orbit of Neptune that is a reservoir for short-period comets

k. small object or particle in space before it enters Earth's atmosphere

l. small region in the sky from which meteor showers appear to originate

m. small object or particle that survives passage through Earth's atmosphere and reaches the ground

Completion

Fill each blank with the most appropriate term from the list for that paragraph. A term may be used once, more than once, or not at all. Check your answers with the Answer Key and review when necessary. If a question requires two or more answers in succession, they may be in any order, unless the question indicates otherwise.

1. At certain times of the year, a patient observer can see flashes of light in the night sky. These flashes of light have been incorrectly called _____. These flashes are produced by small particles or objects and are correctly called _____. The particles themselves are called _____. Periodically, one of the particles is large enough to survive the passage through Earth's atmosphere and reach Earth's surface. Such an object is called a _____.

    Upon careful analysis, these objects can be classified into two groups. The first group is made of metals and is called _____. In fact, these objects are made of _____ and _____. When these objects are cut in half, polished, and etched with _____ acid, regular bands appear. These bands are called _____ patterns. The bands were produced when the object cooled very _____ and can be used in trying to understand the _____ of the object.

    The second group of objects looks very much like Earth rocks. These objects are called _____. This second group can be subdivided into three additional groups. The objects with small, rounded bits of glassy rock imbedded within are known as _____. The bits of glass are called _____. The second subgroup also contain these bits of glass, along with carbon and water. These _____ can exist only if the object has not been heated or melted in any way. This second group, known as _____, contains the least altered remains of the solar nebula. Finally, the objects in the third group contain no glass or _____. They appear to have been subjected to intense _____ and look very similar to Earth _____. This last group is called the _____.

achondrites

carbonaceous chondrites

chondrites

chondrules

cooling

heating

history

iron

iron meteorites

lava

meteorites

meteoroids

meteors

nickel

nitric

rapidly

shooting stars

slowly

stony meteorites

volatiles

Widmanstätten

2. In 1772 a rule predicting the distances to the known planets was announced. Known as the _____ rule, it also predicted the existence of a planet between the orbits of _____ and _____ where one was not known to exist. Although no planet was ever discovered in that region, astronomers discovered many smaller objects. These objects are known as _____. Studies of these objects not in the visible spectrum, but in the _____ spectrum, show that they can also be divided into three groups.

    The S-type objects are very _____ and their color appears _____. They seem to be silicates mixed with _____ and may be associated with meteorites from the _____ group. The C-type objects are very dark and seem to be _____ and associated with the meteorites of the _____ class. A further connection between this group and the meteorites is that these objects are located in the _____ belt, indicating that they have been exposed to extremely _____ temperatures and have experienced very little _____ or melting. Finally, the M-type objects are very bright but without the _____ color of the _____ objects. Available evidence indicates that they are made of metals, probably _____ and _____. These objects are similar to _____.

asteroids

blue

bright

C-type

carbonaceous

carbonaceous chondrites

chondrites

cooler

heating

high

infrared

inner

iron

iron meteorites

Jupiter

low

M-type

Mars

metals

Neptune

nickel

outer

Pluto

red

S-type

Titius-Bode

ultraviolet

3. To early humans, one of the most frightening events occurring in the sky was the appearance of a comet. These objects looked like a fuzzy ball of light with a very long _____. It wasn't until the nineteenth century that science began to understand the nature of these objects. A comet can be described as a mountain of ice and rock. In fact, many call the central lump, or _____, a dirty _____. Vaporization of the ices in this central lump forms the _____. This large cloud of _____ and _____ is what is visible when a comet is viewed through a telescope. Emerging from this vast cloud is the tail of the comet, which may be _____ of kilometers long.

Comets have two kinds of tails. The first is straight and wispy and is a _____ tail. It is made up of _____ gases and it is blown _____ by moving gas and the embedded magnetic field of the solar wind. The second kind of tail is the _____ tail, which is unaffected by magnetic fields. This tail reflects the spectrum of sunlight and is therefore made up of _____ bits of

_____ from the central lump. Although the material in this tail does not respond to magnetic effect, it is pushed _____ by the pressure of _____. It is interesting to note that the tail of a comet is always directed _____ from the sun.

away	millions
billions	nucleus
coma	outward
dust	snowball
gas	solid
ionized	sunlight
liquid	tail

Self-Test

Select the one best answer for each question. Check your answers with the Answer Key and review when necessary.

1. Which of the following are classifications of meteorites?
   a. basalt and igneous.
   b. irons and magnetite.
   c. iron and stony.
   d. shower and crystal.

2. Widmanstätten patterns are found in
   a. achondrites.
   b. iron meteorites.
   c. chondrules.
   d. stony meteorites.

3. Because of the delicate nature of the carbonaceous chondrites, it is believed that these meteorites came from planetesimals that

   a. orbited the sun near the planet Mercury.

   b. collided with Earth billions of years ago.

   c. orbited far from the sun in regions of the solar system that are very cold.

   d. collided with Mars, causing material to be ejected from the planet and eventually land on Earth.

4. The iron meteorites probably formed from

   a. large planetesimals that melted and differentiated to form an iron core.

   b. condensing iron atoms during the cooling period of the early solar nebula.

   c. collisions with Mars-sized planetesimals.

   d. cometary impacts with asteroids.

5. A meteor is

   a. a particle from outer space.

   b. a particle from outer space that lands on Earth.

   c. produced by tiny bits of debris from comets.

   d. a small asteroid.

6. A meteor shower occurs when

   a. an asteroid collides with Earth.

   b. Earth passes through the dust and bits of material left in the orbit of a passing comet.

   c. meteorites are found in a short period of time.

   d. Earth reaches its nearest position to the sun (perihelion).

7. Most of the asteroids revolve about the sun between the orbits of

   a. Jupiter and Saturn.

   b. Mars and Jupiter.

   c. Earth and Mars.

   d. Saturn and Uranus.

8. M-type asteroids appear to
   a. orbit in a fixed position between Jupiter and Mars.
   b. be made of stone and silicates and are red in color.
   c. consist of carbon and look very dark.
   d. be made of iron-nickel alloy and are very bright.

9. The formation of asteroids is the result of
   a. a collision between two planets.
   b. planetary material that never formed into large planets.
   c. an exploded planet.
   d. the escape of Jovian-type satellites.

10. Many collisions caused planetesimals to fragment into smaller pieces. These pieces probably make up the material known as
    a. meteor showers.
    b. meteorites
    c. asteroids.
    d. planetary moons.

11. The solar wind directs the tail of a comet
    a. outward, away from the sun at all times.
    b. away from the sun when the comet enters the solar system and toward the sun when it leaves the solar system.
    c. toward the sun when the comet enters the solar system and away from the sun when it leaves.
    d. toward the sun at all times.

12. The nucleus of the comet contains the ice and rock referred to as a "dirty snowball." The coma is produced when the
    a. nucleus explodes.
    b. comet reaches its most distant point from the sun (aphelion) and the ices vaporize.
    c. tail of the comet compresses back toward the nucleus.
    d. sun causes sufficient heating and the ices of the nucleus vaporize to produce a vast cloud of gas and dust.

13. The source of the long-period comets is a region of space at least 10,000 astronomical units from the sun. This region of space is called

    a. interstellar space.

    b. the interstellar medium.

    c. the Oort cloud.

    d. the Kuiper belt.

14. Studies show that the short-period comets cannot originate at a distance thousands of astronomical units from the sun. A closer source for short-period comets is the

    a. Kuiper belt.

    b. Oort cloud.

    c. Lagrangian points.

    c. asteroid belt.

15. Comet Halley was first observed

    a. in 1910.

    b. more than 2,000 years ago.

    c. during the Norman invasion of A.D. 1066.

    b. by Edmund Halley and Isaac Newton.

16. The Giotto and Vega spacecraft discovered that the nucleus of Comet Halley is

    a. perfectly round and smooth.

    b. bright and irregularly shaped.

    c. very dark and irregularly shaped.

    d. very dark and perfectly smooth, with no obvious relief.

17. Jets issue from the nucleus of Comet Halley only

    a. from the side of the comet pointing away from the sun.

    b. when the comet is at aphelion.

    c. during the period of time when the tail is visible.

    d. from the sunlit side of the comet.

18. Evidence that Earth has collided with an asteroid-sized object is suggested by a layer of the rare element

    a. ytterbium.

    b. plutonium.

    c. iridium.

    d. planetarium.

19. The "smoking gun" crater that fits the description of the asteroid impact that may have caused the extinction of the dinosaurs was found

    a. off the coast of Hawaii.

    b. off the coast of the Yucatán peninsula.

    c. outside of Flagstaff, Arizona.

    d. in Crater Lake, Oregon.

Answer Key

Matching

1.	e	8.	a
2.	d	9.	j
3.	b	10.	l
4.	c	11.	i
5.	g	12.	m
6.	f	13.	k
7.	h		

Completion

1. shooting stars, meteors, meteoroids, meteorite

   iron meteorites, iron, nickel, nitric, Widmanstätten, slowly, history

   stony meteorites, chondrites, chondrules, volatiles, carbonaceous chondrites, volatiles, heating, lava, achondrites

2. Titius-Bode, Mars, Jupiter, asteroids, infrared

   bright, red, metals, chondrites, carbonaceous, carbonaceous chondrites, outer, low, heating, red, S-type, iron, nickel, iron meteorites

3. tail, nucleus, snowball, coma, gas, dust, millions

   gas, ionized, outward, dust, solid, dust, outward, sunlight, away

Self-Test

(Page numbers refer to the textbook.)

   1. c (Objective 1; page 439)

   2. b (Objective 1; page 439)

   3. c (Objective 2; page 442)

   4. a (Objective 2; pages 441–442)

   5. c (Objective 3; page 440)

   6. b (Objective 3; page 441)

   7. b (Objective 4; page 442; video program)

   8. d (Objective 4; page 442)

   9. b (Objective 5; page 443)

  10. c (Objective 5; page 443)

  11. a (Objective 6; page 445)

  12. d (Objective 6; page 445)

  13. c (Objective 7; page 450)

  14. a (Objective 7; Background Notes 25A)

  15. b (Objective 8; Background Notes 25B)

  16. c (Objective 8; page 446)

  17. d (Objective 8; page 447)

  18. c (Objective 9; Background Notes 25D)

  19. b (Objective 9; Background Notes 25D)

L E S S O N
26

Life on Other Worlds

Assignments

For the most effective study of this lesson, we suggest that you complete the
assignments in the sequence listed below.

Before Viewing The Video Program

- Read the Overview and Learning Objectives for this lesson. Use the Learning
  Objectives to guide your reading, viewing, and thinking.

- Read Chapter 20, "Life on Other Worlds," pages 459–482, and Chapter 21,
  "Afterword," pages 483–484, in the textbook.

- Read Background Notes 26, "The Search for Other Planets," following
  Lesson 26 in this guide.

- Read the Viewing Notes in this lesson.

View "Life on Other Worlds" Video Program

After Viewing the Video Program

- Briefly note your answers to questions listed at the end of the Viewing Notes.

- Review all reading assignments for this lesson, especially the textbook
  Chapter 20 summary on page 477, the Background Notes 26, and the Viewing
  Notes in this lesson.

- Write brief answers to all the review questions at the end of textbook Chapter
  20 to be certain you understand the text material.

- Complete the Review Activities in this guide to reinforce your understanding
  of important terms and concepts. Check your answers with the Answer Key at
  the end of this lesson and review when necessary.

- Take the Self-Test in this guide to measure your achievement of the Objectives.
  Check your answers with the Answer Key and review when necessary.

- Complete the Using What You've Learned activities and any other activities
  and projects assigned by your instructor.

Overview

The search for extraterrestrial life is perhaps the greatest challenge presented to science. The search requires the cooperation of several disciplines. Biology is involved because that science defines and describes life as we know it and identifies conditions necessary for life to evolve and be sustained. Astronomy identifies the types of stars that are likely candidates to harbor planets that could support life and astronomers search for planets around those stars. Engineers and other scientists design and build the radio telescopes and computers that survey the sky and search for signals from extraterrestrial civilizations.

Today, most astronomers think the probability of other stars having planetary systems is very strong. With 100 billion stars in the Milky Way galaxy alone, it is almost inconceivable that none would have planets. Of course, the existence of planets does not mean that they have life on them. This lesson explores how astronomers are conducting the search for planets and extraterrestrial intelligence and examines the physical basis of life and how life evolved on Earth.

In the video program, astronomers explain why O, B, A, and M type stars are poor candidates for life-harboring planets. By using our solar system as an example, the experts examine the region around our sun that permits life-bearing planets, consider where the existence of liquid water is possible, and explain why water is so necessary.

Although it is reasonably easy to describe the types of stars that are likely to have life-bearing planets in orbit, detecting a planet around another star is not. A planet would be millions of times fainter than the star it orbits and would be overcome by the starlight if astronomers tried to observe it directly. Detecting a planet as the result of its gravitational influence on the star would also be difficult, because a planet would be at least 100,000,000 times less massive than the star. In spite of these obstacles, efforts are underway to detect planets around other stars.

The most likely approach to successfully detecting an extraterrestrial civilization, however, is to allow radio telescopes and computers to scan the sky, "listening" for signals that would have the signature of an intelligent life form. Even that approach has problems. The radio wavelength is extremely broad, so to what frequencies do we tune our instruments? What messages do we listen for? How do we confirm a positive signal? All these questions have been taken into account by the various projects that fall under the general heading of SETI, the search for extraterrestrial intelligence. These searches are scanning the "magic frequencies" and listening to hundreds of stars each day, hoping to receive some positive indication that planets other than Earth harbor intelligent civilizations.

The likelihood of life existing elsewhere in the galaxy is very strong. The abundance of chemicals needed to support and form life is surprising, and these chemicals are present in the star-forming regions of the galaxy. The rules so far understood about the nature of life are so plausible that it seems inconceivable that these rules would not be in force elsewhere and lead to the development of life. The unknown factor is whether that life has evolved into an intelligence capable of communication. Finally, if ever we do contact extraterrestrial civilizations, we must have an answer for the question, "who will speak for Earth and what would our planet say?"

Learning Objectives

When you have completed all assignments for this lesson, you should be able to:

1. Explain the criteria scientists use to define life. TEXTBOOK PAGES 460–461.

2. Explain the nature of DNA and RNA and how these molecules help to define life. TEXTBOOK PAGES 461–463.

3. Describe the current understanding of how life originated on Earth and the significance of the Miller experiment. TEXTBOOK PAGES 463–467.

4. Describe the characteristics of a star and its solar system that would provide the physical conditions allowing life to evolve. TEXTBOOK PAGES 470–472; VIDEO PROGRAM.

5. Describe the efforts of astronomers to detect planets around other stars. BACKGROUND NOTES 26; VIDEO PROGRAM.

6. Explain how the Drake equation is used to determine the number of possible extraterrestrial civilizations in the galaxy. TEXTBOOK PAGE 476; VIDEO PROGRAM.

7. Describe the recent attempts to communicate with extraterrestrial civilizations and to detect radio transmissions from extraterrestrial civilizations. TEXTBOOK PAGES 472–477; VIDEO PROGRAM.

Viewing Notes

This video program explores one of humankind's oldest questions: Are we alone in the universe?

The first segment of "Life on Other Worlds" identifies the chemicals of life, and astronomers explain how supernovae supplied calcium, iron, and all the other elements that are the basis of life. The video program then examines the role of environment in allowing life forms to develop and to exist. Astronomers describe the early environment of our solar system and the role of comets and asteroids in the origin of life. This segment also looks at the environments on Venus and Mars and explains why they cannot sustain life.

In the segment "Extrasolar Planets," astronomers explain why detection of planets around other stars is so difficult. They describe some of the techniques they do use, including coronagraphs, infrared sensors, and observations to detect wobbles in a star's movement and Doppler shifts in the light from a star. Astronomers also describe the concept of the habitability zone, the area around a sun where water could exist in liquid form, and explain why only certain types of stars are likely candidates for hosting planets with life on them.

The last segment begins with an explanation of the Drake equation, which attempts to estimate the number of technological civilizations in a galaxy. The segment then describes current attempts to search for extraterrestrial intelligence by monitoring radio wavelengths. Astronomers describe the challenges of searching the frequencies spectrum and the use of "magic" frequencies that would be universal.

As you watch the video program, consider the following questions:

1. What are the two most abundant elements in the universe, and where did the heavier elements such as carbon and iron come from and how did they become part of Earth and of us?

2. Why is liquid water an important component in the recipe for life?

3. What role did life itself play in securing its place on planet Earth?

4. What is the habitability zone about a star?

5. Why is it more likely for G and F stars to support life-bearing planets rather than O and A type stars?

6. What are the problems in directly observing or photographing a planet around a distant star?

7. What observations, other than direct detection, can be used to infer the existence of a planet around a distant star?

8. How can observations in the infrared be used to detect the existence of planets around other stars?

9. How does the Drake equation help astronomers determine the potential for extraterrestrial civilizations?

10. What is SETI and how does it use radio waves to listen for extraterrestrial communications?

Review Activities

Matching

Match the terms listed below with the definitions that follow. Check your answers with the Answer Key and review any terms you missed.

F 1. DNA

K 2. protein

b 3. RNA

G 4. enzyme

m 5. amino acid

o 6. natural selection

j 7. mutant

e 8. Cambrian period

n 9. Miller experiment

H 10. primordial soup

d 11. chemical evolution

L 12. stromatolite

i 13. life zone

P 14. water hole

c 15. culture shock

a 16. SETI

a. shorthand reference for the search for extraterrestrial intelligence

b. long carbon-chain molecules that use the information stored in DNA to manufacture complex molecules necessary to the organism

c. the bewildering impact of one society on another

d. chemical process that led to the growth of complex molecules on primitive Earth

e. a geological period 0.6 to 0.5 billion years ago during which life on Earth became diverse and complex. Rocks from this period contain the oldest identifiable fossils

f. long carbon-chain molecule that records information to govern the biological activity of the organism

g. special protein that controls processes in an organism

h. rich solution of organic molecules in Earth's first oceans

i. region around a star within which a planet can have temperatures that permit the existence of liquid water

j. offspring born with altered DNA

k. complex molecule composed of amino acid units

l. structure produced by communities of blue-green algae or bacteria

m. basic building block of protein

n. experiment that reproduced the conditions under which life may have begun on Earth

o. process by which the most adaptive traits are passed on, allowing the most able to survive

p. interval of the radio spectrum between the 21-centimeter neutral hydrogen radiation and the 18-centimeter OH radiation. Likely wavelength to use in the search for extraterrestrial life

Completion

Fill each blank with the most appropriate term from the list for that paragraph. A term may be used once, more than once, or not at all. Check your answers with the answer key and review when necessary. If a question requires two or more answers in succession, they may be in any order, unless the question indicates otherwise.

1. The physical basis for life on Earth is the _____ atom. This atom is capable of forming long, complex, _____ chains that can extract, store, and use _____. These chains are located in the center of the basic unit of life, the _____. Within the _____, these chains take a form that resembles a _____. These complex molecules contain _____ on how the cell _____ and how it _____ as part of the larger organism.

   These complex molecules are called _____, or DNA. The rails of the ladder are made of alternating _____ and _____. Between the rails are _____, which are made of pairs of _____. There are _____ different kinds of _____, the order of which determines how the _____ functions.

   The DNA molecule contains the information necessary to make all the molecules needed by the organism. Segments of the DNA molecule are used in the production of _____, some of which help build the cell wall, while others control body processes such as growth. Those used to control growth are called _____.

   It is not the DNA molecule that actually leaves the cell and begins construction on the _____. The DNA molecule produces a copy of its patterns in a long _____ chain molecule called _____ or RNA. This RNA molecule carries the _____ necessary to use the simple building blocks called _____ and combine them to produce _____. The RNA molecule acts as a _____, carrying out the patterns established by the _____ molecule.

amino acids

bases

carbon

cell

deoxyribonucleic acid

DNA

eight

energy

enzymes

four

functions

information

messenger

nucleus

phosphates

proteins

reproduces

ribonucleic acid

rungs

stable

sugars

twisted ladder

two

2. Recent research seems to indicate that planets around other stars are probably a likely occurrence. However, planets that might harbor life probably occur much less frequently. The first criteria for a planet to harbor life is that its orbit must be _____. It cannot wander too far or too close to the star. To meet this requirement, the planet would probably orbit a _____ star and not a _____ or multiple-star system.

Furthermore, the lifetime of the star must be very long. It probably requires between 0.5 to 1.0 _____ years for life to evolve on a planet. Therefore, the _____ and _____ type stars are unsuitable to harbor life-sustaining planets because their lifetimes are too _____. Other stars that are poor candidates to harbor life-sustaining planets are _____ stars because they are too _____. Planets around such a star would have to orbit too _____ to the star and may become _____ to the star, keeping the _____ side toward the star. In this situation, one side of the planet would be in _____ darkness, causing the atmosphere and its water to _____. Another problem with the _____ stars is that they are subject to sudden _____. This release of a star's energy with the planet being so _____ might destroy any life that had formed.

The most likely candidates for stars that may have suitable planets to harbor life are the _____, _____, and _____ stars. These stars are and their lives are relatively _____ and stable. They also have a large region of space around them where water can remain in _____ form.

This region of space is called a _____, or ecosphere. To summarize, stars that would be suitable to have planets capable of sustaining life are stars that are _____, and of the spectral types _____, _____, and _____. These stars also have _____ lives and have relatively large _____. At present we know of only one star that meets all these requirements and has produced a planet suitable for sustaining life: our sun.

B	K
billion	life zone(s)
binary	liquid
close	long
constant	M
cool	million
cooler	O
different	same
distant	short
F	single
flares	stable
freeze	tidally locked
G	

3. It seems obvious, because of the great _____ between the stars, that physically _____ to the stars is, at least for now, impossible. Therefore, the only form of _____ we will be able to have with extraterrestrial civilizations will be through the use of _____ . These signals have already started leaving Earth in the form of _____ and _____ broadcasts. If any other civilizations pick up these signals, we hope they will still consider us intelligent.

Although scientists may never know if any other civilization has received our signals, either by accident or design, astronomers are _____ for signals reaching Earth from _____ civilizations. Because the spectrum of _____ is so large, it is necessary to narrow the range of _____ to which telescopes can listen. Astronomers have determined that there are some _____ that will probably be used by extraterrestrial civilizations for this kind of communication. These wavelengths involve _____ and _____ that would be familiar and important to all forms of _____. The range of frequencies lies between the

21-centimeter line of _____ and the 18-centimeter line of the _____ molecule. The interval between these two lines has been called the _____ , because the combination of _____ and _____ yields _____ (H_2O). Even if a civilization is not based on _____ , it will certainly know, assuming its beings have an intelligence level at least equal to our own, about the _____ line of neutral hydrogen. Current efforts to detect a signal from an _____ intelligence are called _____. Someday we may have an answer from life on another world.

21-cm	neutral hydrogen
communication	OH
compounds	radio
distances	radio waves
elements	SETI
extraterrestrial	television
hydrogen	traveling
leaking	water
life	water hole
listening	wavelengths
magic frequencies	

Self-Test

Select the one best answer for each question. Check your answers with the Answer Key and review when necessary.

1. The basic unit of life as we know it is the
   a. nucleus.
   b. cell.
   c. DNA.
   d. RNA.

2. The information that enables the organism to survive and is transferred from parent to offspring is found in

   a. DNA.

   b. the cell membrane.

   c. a twisted ladder.

   d. proteins.

3. The information a cell needs to function is found on the DNA molecule in the form of

   a. multiple combinations of unknown chemicals.

   b. RNA subgroups.

   c. a twisted helix.

   d. four bases and the order in which they appear.

4. The information regarding cell functioning is carried out of the nucleus by

   a. amino acids.

   b. RNA.

   c. proteins.

   d. enzymes.

5. Evidence that life began in the oceans has been found in

   a. surface rocks.

   b. sediments on the ocean floor.

   c. Cambrian period fossils.

   d. the primordial soup.

6. The Miller experiment consisted of various gases, believed to have been present in the early atmosphere of Earth, exposed to an electrical arc. This experiment resulted in

   a. nothing.

   b. carbon dioxide gas and water vapor.

   c. amino acids, fatty acids, and urea.

   d. rain.

7. O and A stars are unlikely to harbor planets suitable for sustaining life because

   a. their stable energy-producing life span is too short.

   b. they are too hot.

   c. they are too cold.

   d. they produce an usually high amount of deadly radiation.

8. The region of space around a star in which water can remain in liquid form is called the

   a. water hole.

   b. temperate zone.

   c. main-sequence stage.

   d. life zone.

9. If a large planet orbited a nearby star, astronomers would most likely observe

   a. a faint speck of light very close to the star.

   b. a slight wobble in the star's proper motion.

   c. a red shift in the star's spectrum.

   d. an eclipse as the planet moved in front of the star.

10. As a planet and star orbit about their common center of mass, the center of mass would be observed to

    a. travel across the sky in a straight line.

    b. wobble slightly as the star moved smoothly across the sky.

    c. exhibit blue shift and red shift of the spectral lines as the planet and star orbited.

    d. exhibit only a blue shift in its spectral line as it moved across the sky.

11. The Drake equation is used to

    a. determine the number of stars in the Milky Way galaxy.

    b. count the number of stars visible in the night sky.

    c. determine the spacing distance of planets in our solar system.

    d. estimate the number of technological civilizations within our galaxy.

12. Of the following factors, the one that is a part of the Drake equation is the

   a. temperature of the star during its main sequence stage.

   b. number of O and B stars in the galaxy that exhibit the possible existence of orbiting planets.

   c. fraction of the lifetime of a star during which a technological civilization is capable of communicating.

   d. total number of planets in a solar system orbiting about a binary star.

13. One of the most likely wavelengths of radio energy being used to search for extraterrestrial signals is the

   a. scattered blue light of the interstellar medium.

   b. 21-cm line of neutral hydrogen.

   c. 3 K microwave background energy.

   d. Balmer wavelength of neutral hydrogen.

14. The water hole refers to

   a. a range of radio wavelengths (frequencies) that fall between the wavelengths of neutral hydrogen and the OH molecule.

   b. the region around a star in which water can remain in liquid form.

   c. planets that have been confirmed to contain traces of liquid or frozen water.

   d. nebulae of interstellar matter that have water or OH molecules.

Using What You've Learned

1. Make a detailed list of the reasons O, B, A, and M stars are not expected to shelter intelligent civilizations.

2. Make a detailed list of the reasons F, G, and K stars are the most likely to shelter intelligent civilizations.

3. From Appendix D, page A17, "The Nearest Stars," in the textbook, select the stars nearest the sun most likely to shelter civilizations. Be sure to consider spectral type, luminosity class, and whether they are single or multiple stars systems. Locate these stars on the SC1 Star Chart that you used in your activities in Lesson 21.

Answer Key

Matching

1. f
2. k
3. b
4. g
5. m
6. o
7. j
8. e
9. n
10. h
11. d
12. l
13. i
14. p
15. c
16. a

Completion

1. carbon, stable, energy, cell, nucleus, twisted ladder, information, reproduces, functions

   deoxyribonucleic acid, phosphates, sugars, rungs, bases, four, bases, cell proteins, enzymes

   proteins, carbon, ribonucleic acid, information, amino acids, proteins, messenger, DNA

2. stable, single, binary

   billion, O, B, short, M, cool, close, tidally locked, same, constant, freeze, M, flares, close

   F, G, K, cooler, long, liquid, life zone, single, cooler, F, G, K, long, life zones

3. distances, traveling, communication, radio waves, television, radio

   listening, extraterrestrial, radio waves, wavelengths, magic frequencies, elements, compounds, life, neutral hydrogen, OH, water hole, hydrogen, OH, water, water, 21-cm, extraterrestrial, SETI

Self-Test

(Page numbers refer to the textbook.)

1. b (Objective 1; page 461)

2. a (Objective 1; pages 461–462)

3. d (Objective 2; page 461)

4. b (Objective 2; page 462)

5. c (Objective 3; pages 463–464)

6. c (Objective 3; page 464)

7. a (Objective 4; page 471; video program)

8. d (Objective 4; page 472; video program)

9. b (Objective 5; Background Notes 26; video program)

10. a (Objective 5; Background Notes 26; video program)

11. d (Objective 6; page 476; video program)

12. c (Objective 6; page 476; video program)

13. b (Objective 7; page 474; video program)

14. a (Objective 7; page 474)

Unit IV

Background Notes

Radar Astronomy

In current astronomical research, radio astronomy has been invaluable in probing the far reaches of the universe. A specialized type of radio astronomy, **radar astronomy**, has been equally valuable in revealing the secrets of our solar system's planets. Radar astronomy is especially useful for studying planets that are difficult to observe through optical telescopes. Mercury, for example, is too close to the sun for optical study of its surface, and Venus is hidden behind a thick cloud cover.

Radar stands for **ra**dio **d**etecting **a**nd **r**anging. In radar astronomy, transmitters send artificially produced radio waves toward the object under study. Analysis of the reflected signal can provide information about a planet's distance from Earth, its rotation rate, and its surface features. To date, the best-known applications of radar in astronomy have been mapping of the surface of Venus and measurement of the rotation rates of Mercury and Venus.

Determination of a planet's distance is made by measuring the time it takes the radio signal to travel to the planet and return to a receiving dish on Earth. If the returning radio "echo" is Doppler-shifted in frequency or wavelength, the speed of the planet in orbit around the sun can be deduced. In the case of Mercury and Venus, this information was used to determine the value of the astronomical unit in miles or kilometers.

Because the planet is rotating, the reflected signal (the original signal was a single frequency or wavelength) has a spread of wavelengths. This spread in wavelength is produced by the radio signal being reflected back from different parts of the planet. A portion of the radio signal is reflected from the part of the planet rotating toward Earth, which causes the signal to be slightly Doppler-shifted toward higher frequencies, or shorter wavelengths. A portion of the radio signal is also reflected off the part of the planet rotating away from Earth, which causes the reflected signal to be Doppler-shifted toward lower frequencies, or longer wavelengths. The spread in wavelengths is used, along with measurements of speed and diameter, to determine the rotation rate of the planet.

Since the planet is spherical, but not perfectly smooth, the reflected signal is not detected back at Earth or in a spacecraft all at once. Because tall mountains and the bottoms of craters are different distances from the transmitter, the reflected signal from these surface features reaches the receiver at different times. By keeping track of the time between signal transmission and reception, as well as any Doppler shift in the signal, a unique point on the planet's surface can be mapped. A computer then compiles all the points together to reveal a picture of the surface of a planet.

Although radio waves are longer than light waves and cannot resolve the smallest of surface features, the Magellan spacecraft was able to create images of features on Venus as small as 100 meters wide, a great improvement over the 100-kilometer resolution capability of the first radar-astronomy images. Radar astronomers are working on methods to improve the resolution to capture even smaller details.

Additional information about the surface of the planet can be acquired by analyzing the intensity of the reflected radar signal. Because radio waves can penetrate the surface of the planet, the material in the surface affects the signal's intensity, and the reflection provides information about the composition and texture of the surface. This type of analysis detected the water ice at the poles of Mercury. This ice, located at the bottom of craters that are permanently hidden from sunlight, is also covered by dark insulating material and cannot be observed in optical wavelengths. Analysis of the radar signal indicates the bright polar regions could only be produced by the existence of water ice.

You should note that the colors in the radar images you see in the video programs and reproduced in the textbook are artificially added and do not represent the colors one would see on the surface of the planet. Because the human eye is not sensitive to radio wavelength radiation, any image created using radar must be translated, through the use of "false" colors, into an image that the human eye can see. For instance, colors in the Pioneer radar images identify different levels of elevation. The Magellan images of Venus are similar in color to the images returned by the Soviet Venera spacecraft, which photographed the surface in natural light.

Synchronous Rotation

If you were to stand on the surface of the near side of the moon looking toward Earth, you would begin to notice something very strange. Earth would continue to remain high in the sky throughout the entire time that the moon orbited Earth. Earth would never set. Earth appears nearly "fixed" in the sky because the rate at which the moon rotates on its axis is equal to the length of its orbital period around Earth. This phenomenon is referred to as **synchronous rotation**.

In the video program, Bruce Murray, a planetary scientist at California Institute of Technology, explains that the moon is asymmetrical, not perfectly spherical or uniformly dense. This asymmetry causes Earth to exert tidal forces (differences in the gravitational force due to different distances from Earth). These forces eventually slowed the moon's rotation rate until it became locked onto the Earth, and its rotation rate became equal to its revolution period.

Mercury is also locked into a peculiar rotation-revolution relationship with the sun. Mercury's orbit is highly eccentric (only the orbit of Pluto is more elliptical).

Observations and orbit calculations led astronomers to believe that one side of Mercury faced the sun at all times because of tidal locking, similar to what is in effect with the moon and Earth.

In 1965, however, radar studies using the Arecibo telescope showed that Mercury has a rotation period of 59 days as compared to an orbital period of 88 days. These figures mean that Mercury rotates three times on its axis as it orbits the sun twice. On Mercury, one would observe the sun crossing the meridian (noon) once every 176 days. The length of a solar day on Mercury is equivalent to 176 Earth days. The length of Mercury's day creates some difficulty in studying the surface of the entire planet, even from space. For instance, the Mariner spacecraft flew by Mercury three times and, each time, the same side of Mercury was facing the spacecraft.

BACKGROUND NOTES 25A

Kuiper Belt

The *Horizons* textbook suggests the possibility of the existence of debris that is the source of the short-period comets (comets with low orbital inclination and periods of revolution from only a few years to tens of years) that is not associated with the well-known Oort cloud. This new source of cometary material, now known as the Kuiper belt, or disk, was first suggested by Gerard Kuiper in 1951. He hypothesized that the solar nebula did not end at the distance of Neptune, but rather extended beyond Neptune and that there should be a belt or disk of leftover material that would be available for forming additional planets.

Further support for the existence of such a belt was provided in 1983 when the Infrared Astronomical Satellite detected disks around the nearby stars Beta Pictoris and Vega. Theoretical calculations in the late 1980s showed that short-period comets required the existence of the Kuiper belt in order to account for the inclination of their orbits and their revolution period around the sun. The only missing component needed to verify the existence of the belt was to find objects located within the Kuiper belt region.

In 1977 the first of those objects was found when Charles Kowal, an astronomer at Palomar Mountain Observatory, discovered a new object. Scientists originally thought the object was a stray asteroid orbiting the sun at a distance that ranged from within the orbit of Saturn to beyond the orbit of Uranus. The object did not belong to the main belt of asteroids between Mars and Jupiter. The Minor Planet Center labeled the object Chiron. Measurements of its brightness, based on the standard model of a comet, indicated that Chiron had a diameter of between 130 and 400 kilometers. Comets have diameters of 3 to 10 kilometers. This object was certainly not a comet.

It wasn't until early 1988 that some of the mystery surrounding Chiron was solved. The object began to brighten significantly, providing astronomers with an opportunity to speculate about the nature of the surface of Chiron. The consensus was that the surface of Chiron consisted of ices that could vaporize at very low temperatures, such as carbon monoxide, methane, and nitrogen. (At those distances, it is too cold for water ice to vaporize.) Based on these observations and additional calculations, astronomers concluded that Chiron could not have been in its current orbit for the entire age of the solar system. This conclusion means that Chiron must have come from somewhere else in the solar system and that its original orbit was disturbed by the gravitational influences of Jupiter and Saturn. Further studies suggested that the surface of Chiron more closely resembles the surfaces of Triton and Pluto than the surface of a comet.

In August 1992 Dave Jewitt and Jane Luu of the University of Hawaii discovered a slow-moving object determined to be at a distance of 44 astronomical units. Its diameter was determined to be between 200 and 250 kilometers. Six months later another similar object was discovered, and six months later another two objects were discovered. These objects of ice and rock, at distances of between 35 to 45 astronomical units, have diameters between 50 and 500 kilometers. They are similar in size and brightness to Chiron. Based on the amount of sky that has been searched to date, astronomers estimate that 1,000 to 10,000 of these objects are within the Kuiper belt region. These objects are the largest members of the Kuiper belt and therefore have been the "easiest" to discover. There are always many more small objects than large objects. If our current understanding is correct, the nucleus of a typical short-period comet should be 10 to 20 kilometers wide—small compared to the larger Chiron-type objects already discovered. As a result, there may be 100 million comets within the Kuiper belt—comets that potentially could be disturbed into a trajectory that would bring them into the inner solar system.

Comet Halley—History

The first recorded observation of the celestial object we know as Comet Halley was in 240 B.C. Throughout much of Western civilization and history, comets, especially Comet Halley, have been linked to catastrophic events. This association may have started with Aristotle, who noted the appearance of comets before and after severe storms. Although the storms and the comets were not related, Aristotle's speculations, supported by his reputation and stature, may have given rise to superstitions that connected comets with war, disease, pestilence, famine, and especially the deaths of kings and nobility.

For instance, comets, have often been associated with the destruction of city states and the overthrow of kings. In the first century A.D., Josephus, the Jewish historian, described an omen that foretold the destruction of Jerusalem: a bright comet that "hung like a sword." Comet Halley was observed in A.D. 66, and the Romans attacked Jerusalem in A.D. 69. One should, however, be careful about confusing a prediction with a connection. Josephus wrote about the comet and the fall of Jerusalem after the fact. Other historically significant events that have been associated with apparitions of Comet Halley include the defeat of Attila the Hun in 451 A.D., a plague that spread across Europe in A.D. 530, and the eruption of Mt. Vesuvius in A.D. 684.

The most famous apparition of Comet Halley in European history was associated with the Battle of Hastings in A.D. 1066. This battle occurred when Duke William of Normandy invaded and conquered England. William may have observed Comet Halley in the western sky in the spring of 1066, and this observation may have motivated him to invade England in the fall of that year. A depiction of the Battle of Hastings and the return of Comet Halley has been embroidered into the Bayeux Tapestry. This 231-foot-long medieval tapestry consists of 72 scenes that illustrate the Norman conquest of England.

Although barely mentioned in European records, the most impressive and spectacular apparition of Comet Halley occurred in A.D. 837 and was carefully recorded by Chinese astronomers. In March of that year, the tail of the comet was more than 20 times the diameter of the full moon. Within three weeks, the tail stretched across nearly one-third of the entire sky and had split into two sections. Modern calculations have since determined that the comet the Chinese observed came within 5 million kilometers of Earth. During its closest approach, Comet Halley traversed one-fourth of the entire sky in a period of one day. At its maximum, it was nearly as bright as Venus, and the tail lengthened to $76°$, or nearly half the sky. It was during this apparition that Chinese astronomers concluded that the tail of a comet always pointed away from the sun, a fact that would take Western astronomers nearly 700 years to rediscover.

Although the most spectacular apparition occurred in A.D. 837, the three most scientifically significant apparitions occurred in A.D. 1531, 1607, and 1682. Those apparitions were neither the brightest nor the closest, but they were the three recorded encounters that Edmund Halley used to calculate the orbit of the comet and predict its return in A.D. 1758. These calculations, based on Isaac Newton's law of universal gravitation, were the first demonstration that the laws of physics discovered on Earth were applicable to objects millions of kilometers away.

Superstitions about Comet Halley resurfaced at the time of its appearance early in the twentieth century. Even though astronomers had identified some of the gases in the tail of the comet and explained that the tail was very thin and would cause no problems on Earth, the announcement that Earth would pass through the tail of Comet Halley in 1910 caused many people to panic and a number of people to go into business. Comet pills were very popular in the southeastern United States, and breathing masks were also in demand. It was also rumored that if one stood in a bucket of rainwater while the comet was overhead, the gases in the tail would be dissolved by the water and the individual would remain unharmed. Even the American author Mark Twain used Comet Halley to make a prediction about his own life. He said in 1909, "I came in with Halley's Comet in 1835. It is coming again next year, and I expect to go out with it. It will be the greatest disappointment of my life if I don't go out with Halley's Comet. . . . Oh! I am looking forward to that!" Mark Twain did go out with Comet Halley in 1910. Superstitions are revived with each apparition of Comet Halley and will probably continue into the future. The next apparition of Comet Halley will be in July of 2061.

Comet Shoemaker-Levy 9

In March 1993 astronomers Gene and Carolyn Shoemaker and David H. Levy announced the discovery of a comet. Although many comets are discovered each year by professional as well as amateur astronomers, this comet was special. First, it was one of the rare comets that was in orbit about Jupiter instead of the sun. Orbital calculations showed that the comet had been captured by Jupiter's gravitational influence more than a century ago. Second, Comet Shoemaker-Levy 9, as it was named, had broken into 20 fragments. Orbital calculations again showed that the disintegration of the comet took place during its closest approach to Jupiter in July of 1992. It was clear to astronomers that the comet was not strongly put together because the tidal forces exerted by Jupiter on the comet were not very great. Finally, and most important, Comet Shoemaker-Levy 9 was on a collision course with Jupiter itself, scheduled to impact the planet in July of 1994.

After spreading out over the two years from breakup to impact, the 20 fragments approached Jupiter in a thin line. In fact, astronomers calculated that it would take five and a half days for all the fragments to strike the planet. In predicting the consequences of the impacts, astronomers faced several problems. The first, and probably the most difficult to resolve, was to estimate the size of each piece. At a distance of 480 million miles from Earth, the comet had a halo of gas and dust around it that made it difficult to determine the size of each piece. Estimates of the size of the pieces ranged from 200 feet (60 meters) to 2 miles (5 kilometers). The second problem was to determine how deeply into the Jovian atmosphere the fragments would penetrate. The pieces, traveling at 130,000 miles an hour, were either going to explode upon impact with the atmosphere or penetrate deep into the clouds, causing material from Jupiter to erupt into a giant plume.

When the fragments, labeled A through W, finally struck Jupiter, they released more energy than if all the nuclear weapons on Earth were to be detonated at once. The first and smallest fragment produced a fireball 4,000 kilometers across and a plume 1,000 kilometers high. The energy released was equal to more than a million times the energy released by the A-bomb dropped on Hiroshima, Japan, at the end of World War II. Although the impact sites were on the far side of Jupiter, the planet's rapid rotation brought the impact sites into view of Earth telescopes within a half hour. Fortunately, however, the Galileo spacecraft was in a position to photograph the moment of impact for some of the fragments.

The largest fragment, labeled G, was estimated to be 5 kilometers (2 miles) across. It struck Jupiter with a force equal to 6 million megatons of TNT. The blast released energy equal to that of 6 million million tons of dynamite, all in just a few minutes. In comparison, the United States was estimated to have 20,000 megatons and the Soviet Union 60,000 megatons of explosive power during the height of the Cold War. Although the blast from fragment G created a plume of material 2,000 kilometers high, there was no evidence, from spectroscopic studies, of water in the plume. Astronomers concluded that the explosion was relatively shallow and no material from deep within Jupiter was ejected. The impact of G and the other largest pieces also produced dark spots thousands of kilometers across that were easily viewed from Earth-based telescopes. Preliminary results indicate that these dark spots are composed of material from the comet itself and not material from Jupiter's interior.

Observatories throughout the world focused on this display. Viewing the Comet Shoemaker-Levy 9 impacts in visible as well as infrared light, astronomers witnessed an event that has been repeated countless times since the formation of the solar system.

Interplanetary Debris Impacts

Comet Shoemaker-Levy 9 was a spectacular instance of interplanetary debris hitting a planet. Such impacts have occurred throughout the solar system, including Earth. On Earth, most evidence of impacts has been obliterated through erosion and the effects of plate tectonics. However, some evidence of such impacts from either comets or asteroids has been found in several places, including near the Yucatán Peninsula in Mexico and in Siberia. Also, military records contain data about possible impacts of space debris on Earth's surface or in Earth's atmosphere.

Extinction of the Dinosaurs and the Yucatán Peninsula

In 1980 Luis and Walter Alvarez of the University of California at Berkeley proposed that a large meteorite or cometary impact, occurring at the end of the Cretaceous period, was responsible for the mass extinction of the dinosaurs and many other species. They based their conclusion on the existence of the element iridium in a layer of Earth's crust dating 65 million years ago. Iridium is a rare element on Earth, but thousands of times more abundant in primitive meteorites. This iridium-rich layer has been found at more than 100 locations throughout the world.

If the Alvarezes are correct, a 10-kilometer object, either an asteroid or a comet traveling at a speed of 15 kilometers a second, collided with Earth. This object was three times larger than the largest fragment of Shoemaker-Levy 9. The impact ejected millions of tons of dust into the air, eliminating or significantly reducing the amount of light reaching the surface of Earth. Fires, noxious fumes, and acid rain would also have accompanied the impact. In addition to the biological consequences, the asteroid/comet would also have left a crater 150 to 200 kilometers wide.

Since the Alvarezes first proposed their hypothesis, many scientists searched for that crater.

The crater may have been discovered in 1978 by the Mexican national oil company, Pemex. Geologists conducting a survey of magnetic anomalies off the coast of the state of Yucatán discovered a circular feature resembling an impact crater. The crater, buried under water, rock, and soil, was very difficult to decipher. (Details of the discovery process of this crater can be found in an article in the July 1991 issue of *Astronomy*.) Although other craters are being considered as possible sites for an impact causing the mass extinctions 65 million years ago, many planetary scientists accept the hypothesis that such an impact did cause the end of the reign of the dinosaurs. The current question concerns the possibilities of a smaller or similar impact occurring on Earth in the near future.

Siberian Explosion and Other Evidence

In 1908 a large explosion occurred over the Tunguska region of Siberia. Witnesses described a fiery tail striking Earth. A man, 60 miles away, was knocked down, and shock waves were recorded in England. In Siberia a forest roughly the size of Rhode Island was leveled. (Evaluations of the damage done and the fact that no crater exists indicate that the object, either a comet or an asteroid, exploded 4 miles above Earth's surface.)

In addition to the evidence from Siberia, the U.S. military has kept classified records of high-altitude impacts and collisions with Earth's surface. These records show that, on the average, there are eight explosions a year in the upper atmosphere caused by meteors. These explosions are equal in magnitude to a small atomic bomb. One explosion, believed to be an asteroid, occurred in the Pacific in

1978 and was equal to 100 kilotons of TNT. In 1991 an asteroid, estimated to be 30 feet in diameter, passed between Earth and the moon. A 2,200-pound meteorite exploded over the city of Mbale, Uganda, in 1992. Fragments damaged a train station and cotton factory. One boy was struck in the head by a fragment. He survived.

Because of its larger size, Earth has been struck more often by asteroids and meteorites than has the moon. It is anticipated that Earth will be struck again in the future. Planetary scientists are searching the skies to try to predict if Earth is in any immediate danger. Others are trying to decide what to do in case an object is discovered to be on a collision course. The search for objects will continue, along with the search for a solution to the problem.

BACKGROUND NOTES 26

The Search for Other Planets

Detection of planets around other stars is extremely difficult. At present, it is impossible to see directly a planet around another star. No telescopes on Earth or in orbit above Earth can separate a planet and its star into two images. The star is too bright and the planet too faint. At some observatories, however, astronomers use a coronagraph to block out the light of the star, producing an artificial eclipse. By blocking out the bright light of the star, astronomers may be able to observe the fainter planet.

Other astronomers are trying to detect a planet in orbit about a star by looking for evidence of its gravitational effects. If no planets are in orbit about a star, the star will move across the sky in a straight line. In other words, the star's proper motion would be in a straight line. However, if a planet does orbit a star, the planet and star orbit around a common point called the center of mass. This point is the same center that two stars in a binary system orbit as described in textbook Chapter 8, page 164, and in Figure 8-11. (The center of mass would be much closer to the star than to the planet, because the star is much more massive.) In this case, the center of mass of the planet-star system would move across the sky in a straight line, and the star's path would appear to wobble slightly as it moved across the sky. The planet and star tug on each other gravitationally, and this tug-of-war is revealed in the wobbling motion of the star. Such an observation would require decades of very careful measurement to confirm the existence of a planet.

Another possible method of discovering a planet around another star is to look for a Doppler shift in the light of the star. This shift in wavelength of the light would be due to the wobbling the planet would cause as the star and planet orbited the common center of mass. This effect is similar to the observations made of spectroscopic binaries as described in Chapter 8, pages 166-167, and in Figure 8-14. The effect is more difficult to detect because the wobble again is not as extreme as in a binary star system because a planet has so much less mass than a star.

Finally, some astronomers are searching for planets with instruments that detect infrared energy. Stars are much brighter in visible light than planets. They may be as much as a billion times brighter. In the infrared, however, stars are only 10,000 to 100,000 times brighter than planets. Infrared measurements may prove to be a very useful approach to detecting the presence of planets around other stars.